Conservative Systems and Quantum Chaos

DATE DUE

JUN 2 3 2003	
242562	

DEMCO, INC. 38-2971

Other Titles in This Series

FIELDS INSTITUTE COMMUNICATIONS

THE FIELDS INSTITUTE FOR RESEARCH IN MATHEMATICAL SCIENCES

Conservative Systems and Quantum Chaos

Larry M. Bates
David L. Rod
Editors

American Mathematical Society
Providence, Rhode Island

The Fields Institute for Research in Mathematical Sciences

The Fields Institute is named in honour of the Canadian mathematician John Charles Fields (1863–1932). Fields was a remarkable man who received many honours for his scientific work, including election to the Royal Society of Canada in 1909 and to the Royal Society of London in 1913. Among other accomplishments in the service of the international mathematics community, Fields was responsible for establishing the world's most prestigious prize for mathematics research—the Fields Medal.

The Fields Institute for Research in Mathematical Sciences is supported by grants from the Ontario Ministry of Education and Training and the Natural Sciences and Engineering Research Council of Canada. The Institute is sponsored by McMaster University, the University of Toronto, and the University of Waterloo and has affiliated universities in Ontario and across Canada.

1991 *Mathematics Subject Classification*. Primary 58F05;
Secondary 81–06.

Library of Congress Cataloging-in-Publication Data

Conservative systems and quantum chaos / Larry M. Bates, David L. Rod, editors.
p. cm. —(Fields Institute communications, ISSN 1069-5265; vol. 8)
Includes bibliographical references.
ISBN 0-8218-0254-2
1. Quantum chaos—Congresses. 2. Hamiltonian systems—Congresses. I. Bates, Larry M. (Larry Meredith), 1957– . II. Rod, David L. (David Lawrence), 1938– . III. Workshop on Conservative Systems and Quantum Chaos (1992 : Waterloo, Ont.) IV. Series: Fields Institute communications ; v. 8.
QC174.17.C45C66 1996
530.1′2–dc20 96-1821
CIP

---

10 9 8 7 6 5 4 3 2 1 01 00 99 98 97 96

Contents

Preface

This volume presents new research in classical Hamiltonian and quantum systems from the Workshop on Conservative Systems and Quantum Chaos held during The Fields Institute Program Year on Dynamical Systems and Bifurcation Theory in October 1992 in Waterloo, Canada.

The workshop was organized so that there would be a number of presentations bridging the gap between classical and quantum mechanical systems. Four of these papers appear in this collection:

- M.S. Alber and J.E. Marsden apply a previously developed method based on 'asymptotic reduction' for obtaining geometric phase phenomena from soliton equations to show, in particular, that one can 'hear' the semiclassical monodromy of the spherical pendulum.
- T.J. Bridges uses a variational principle to study the linear stability of periodic plane waves of nonlinear Hamiltonian evolution equations with applications to the nonlinear Schrödinger equation.
- M. Kummer presents 'a unified approach to the theory of nonlinearly perturbed harmonic oscillators in classical and quantum mechanics' with applications to the quantum mechanical Kepler problem.
- K. Meyer, J. Delos and J.-M. Mao find generic structures of bifurcations for periodic orbits of two-parameter area-preserving maps that compare with experimental results of atomic spectra.

The remaining six papers in this volume cover a variety of topics in Hamiltonian dynamics:

- E.A. Lacomba and J.G. Reyes use blow-up techniques to describe the global flow on a closed half-strip of the Coulombian restricted 3-body problem.
- E. Lerman and R. Sjamaar give conditions under which the action of a reductive complex Lie group on a Kähler manifold can be linearized in the neighborhood of a fixed point.

- D.C. Offin and H.-F. Yu use global variational arguments to establish the existence of infinitely many distinct hyperbolic periodic orbits in the forced pendulum equation.
- G.W. Patrick numerically investigates the drift of the rotation vector for a mechanical system consisting of two rigid bodies coupled by an ideal ball and socket joint.
- B.D. Sleeman studies the spatio-temporal chaos that can arise in certain biological models from, in particular, population genetics and ecology.
- J.C. van der Meer analyzes the degenerate Hamiltonian Hopf bifurcation building upon his previous (1985) work on the non-degenerate case.

The editors of this volume wish to thank Dr. Richard Cushman for help in organizing the workshop and many other tasks above and beyond the call of duty.

Fields Institute Communications
Volume 8, 1996

Semiclassical Monodromy and the Spherical Pendulum as a Complex Hamiltonian System

Mark S. Alber
Department of Mathematics
University of Notre Dame
Notre Dame, IN 46556

Jerrold E. Marsden
Control and Dynamical Systems 104-44
California Institute of Technology
Pasadena, CA 91125

1 Introduction

In Alber and Marsden [1992, 1994a, 1994b], we introduced a new method for obtaining geometric phase phenomena for soliton equations, including the familiar phase shift of interacting solitons—see for example, Ablowitz and Segur [1981]. Prior to this work, the phase spaces of integrable systems were viewed as being foliated by invariant tori; however, for soliton phase spaces and to get formulas for the geometric phases, we have shown that a foliation by noncompact varieties is essential. The method is based on what we called *asymptotic reduction*. This procedure is applied to a new complex angle representation on a noncompact invariant variety and it leads to a description of geometric phases in terms of the monodromy of the phase function at singularities on hyperelliptic Jacobian fibrations. Monodromy and action-angle variables for Hamiltonian systems were investigated in Duistermaat [1980], Cushman and Duistermaat [1988], Ercolani [1989], Baider, Churchill and Rod [1990], Alber and Marsden [1994a], Bates and Zou [1993], amongst others.

In the present paper, we apply our approach to effects that are related to the problem of semiclassical monodromy. A crucial point in doing this is to consider the complexification of the system. In particular, we describe new complex angle representations and Hamiltonians for the classical simple spherical pendulum (our angle representations hold for the n-dimensional case and are different from those of the above mentioned authors even in the real case, and are based on the Abel–Jacobi map). In particular, this yields new exponential Hamiltonians and angle representations on homoclinic varieties and leads to the introduction of Maslov indices of closed curves in Lagrangian submanifolds of the cotangent bundle of the configuration space. These Lagrangian submanifolds are defined by the first

1991 *Mathematics Subject Classification.* Primary: 58F05, 58F07; Secondary: 70H05, 81S99.

integrals of the problem and will be described below. Then we develop complex geometric asymptotics with corresponding quantum conditions. These quantum conditions include classical and complex monodromy together with phase shifts that are related to Maslov indices after transporting the system along certain closed curves in the space of parameters. We refer to these types of phase shifts which are associated to the quantum conditions and also phase shifts that are associated to singularities in the space of parameters as *semiclassical monodromy*.

Cushman and Duistermaat [1988] used Bohr-Sommerfeld quantum conditions in the 2-dimensional case (i.e., a pendulum moving in 3-space) and, using a numerical investigation, detected monodromy in the semiclassical spectrum of the Schrödinger operator. They also pointed out that the main difficulty is related to the construction of action-angle variables at the singular points. Guillemin and Uribe [1989] suggested that one should relate monodromy to Maslov effects and asked if one can "hear" monodromy for the spherical pendulum. In this paper, we establish this relation and show that indeed one can "hear" the semiclassical monodromy (in the sense that one can hear similar semiclassical modes in the acoustic problem).

To carry out this task, we obtain exact formulae for complex modes, the main ingredients in geometric asymptotics, which are defined on the covering space of the Riemann surface under consideration. (These modes can be treated in a way similar to the case of acoustic modes in the whispering gallery phenomenon. For a detailed description of this phenomenon see Keller and Rubinow [1960] and Alber [1989, 1991]). Then we obtain quantum conditions of Bohr-Sommerfeld-Keller (BSK) type, which include Maslov indices. We investigate these quantum conditions on the bundle of Riemann surfaces over the base space consisting of the spectral data, considered as parameters of the problem. Finally, we describe semiclassical monodromy which has contributions from classical (real) monodromy and complex monodromy, as shifts in the quantum conditions resulting from the transport of a semiclassical mode around certain closed curves in the space of parameters.

2 Monodromy and Complex Angle Representations

In this section we obtain complex Hamiltonians for the spherical pendulum. The method works in the general n-dimensional case and is presented in the Appendix; in the main text we focus on the two dimensional case (that is, the ordinary spherical pendulum in three-space) and just specialize the relevant results from the appendix to this case.

Our approach resolves the issue of multivaluedness for the angle representations through the introduction of Riemann surfaces. The associated Riemann surfaces depend parametrically on values of the first integrals of the problem and have singularities, as described below; these manifolds are Lagrangian submanifolds away from singularities. If one fixes values of the first integrals, then for each closed curve in phase space (normally chosen to be in the real phase space and passing through these Lagrangian submanifolds), one can define an associated Maslov index. This index plays a crucial role in the semiclassical geometric asymptotics as we shall see below.

As we shall see, the main difference between the spherical pendulum and the problem of geodesics on a sphere is the presence of additional singularities of these Lagrangian submanifolds. We will show that the angle representation in this case is similar to that for umbilic geodesics (see Alber, Camassa, Holm and Marsden [1995]).

2.1 Some Complex Geometry for the 2-dimensional Spherical Pendulum The Hamiltonian for the spherical pendulum in Cartesian coordinates Q_j, $j = 1, 2, 3$, and their conjugate momenta P_j has the form

$$H = \frac{1}{2}\left(\sum_{j=1}^{3} P_j^2 - \left(\sum_{j=1}^{3} P_j Q_j\right)^2\right) + Q_3. \tag{2.2.1}$$

Here the acceleration due to gravity and constraint of the length of Q have been set equal to one. In this Hamiltonian, the constraint

$$\sum_{j=1}^{3} Q_j^2 = 1 \tag{2.2.2}$$

has been enforced by the extra term

$$\left(\sum_{j=1}^{3} P_j Q_j\right)^2 .$$

Hamilton's equations on $\mathbb{R}^6$ for (2.2.1) *automatically preserve the set* (2.2.2). Notice also that this extra term vanishes on the constrained phase space, so the Hamiltonian becomes the usual expression of kinetic energy plus potential energy on this constrained phase space. This construction, which also holds in the n-dimensional case, provides an example of a Hamiltonian that automatically preserves a desired set of constraints.

Now we change to new variables z_1 and z_2 and their conjugate momenta P_{z_1}, P_{z_2} as follows. Let θ_1 and θ_2 be spherical coordinates, so that $Q_1 = \cos\theta_1 \sin\theta_2$, $Q_2 = \sin\theta_1 \sin\theta_2$, and $Q_3 = \cos\theta_2$, and let P_{θ_1} and P_{θ_2} be their conjugate momenta. Let

$$\begin{aligned} z_1 &= \cos^2\theta_1, \\ z_2 &= \cos\theta_2, \\ P_{z_1}^2(1 - z_1)z_2 &= P_{\theta_1}^2, \\ P_{z_2}^2(1 - z_2^2) &= P_{\theta_2}^2. \end{aligned} \tag{2.2.3}$$

This change of variables is canonical and in terms of them, the Hamiltonian becomes

$$H = \frac{1}{2(1 - z_2^2)} P_{z_1}^2 (1 - z_1) z_1 + \frac{1}{2} P_{z_2}^2 (1 - z_2^2) + z_2. \tag{2.2.4}$$

The n-dimensional version is given in the appendix.

Spherical coordinates provide a convenient "nested" structure that leads to the first integrals, whereas the variables z_j provide variables on the Riemann surfaces determined by the first integrals. We will demonstrate this below explicitly for the spherical pendulum.

2.2 The Angle Representation in the 2-dimensional Case First integrals for the spherical pendulum are the angular momentum about the z-axis and the energy, which we denote β_1 and β_2. A computation establishes the identities

$$
\begin{aligned}
P_{z_1}^2 &= \frac{\beta_1^2}{(1-z_1)z_1}, \\
P_{z_2}^2 &= 2\frac{\beta_2^2}{(1-z_2^2)} - \frac{\beta_1^2}{(1-z_2^2)^2} - 2\frac{z_2}{(1-z_2^2)} \\
&= \frac{2(\beta_2^2 - z_2)(1-z_2^2) - \beta_1^2}{(1-z_2^2)^2}.
\end{aligned}
\tag{2.2.5}
$$

This yields the following expression for the action function

$$
\begin{aligned}
S &= S_1(z_1) + S_2(z_2) \\
&= \int_{z_1^0}^{z_1} \sqrt{\frac{\beta_1^2}{(1-z_1)z_1}}\, dz_1 + \int_{z_2^0}^{z_2} \sqrt{\frac{M(z_2)}{(1-z_2^2)^2}}\, dz_2,
\end{aligned}
\tag{2.2.6}
$$

which will be regarded as the generating function of a canonical transformation from coordinates (z_j, P_{z_j}) to (β_k, α_k) to be described below. Here

$$
M(z) = 2\left(\beta_2^2 - z\right)\left(1 - z^2\right) - \beta_1^2 \tag{2.2.7}
$$

is a basic polynomial of the hyperelliptic curve (Riemann surface) given by

$$
\begin{aligned}
W^2 = M(z) &= 2\left(\beta_2^2 - z\right)\left(1 - z^2\right) - \beta_1^2 \\
&= (z - m_1)(z - m_2)(z - m_3),
\end{aligned}
\tag{2.2.8}
$$

where, $-1 < m_3 < 0 < m_2 < 1 < m_1$. We choose β_1 and β_2 as action variables of the problem and introduce the following conjugate variables

$$
\begin{aligned}
\alpha_1 &= -\frac{\partial S}{\partial \beta_1} = -\int_{z_1^0}^{z_1} \frac{1}{\sqrt{(1-z_1)z_1}}\, dz_1 + \int_{z_2^0}^{z_2} \frac{\beta_1 dz_2}{(1-z_2^2)\sqrt{M(z_2)}} = \alpha_1^0, \\
\alpha_2 &= -\frac{\partial S}{\partial \beta_2} = -2\beta_2 \int_{z_2^0}^{z_2} \frac{dz_2}{\sqrt{M(z_2)}} = t + \alpha_2^0.
\end{aligned}
\tag{2.2.9}
$$

Here $\alpha^0 = (\alpha_1^0, \alpha_2^0)$ is the base point of the angle representation $\alpha = (\alpha_1, \alpha_2)$ and t is time in the original Hamiltonian system. In the real case, z_1 and z_2 move along cycles l_1 and l_2 over the basic cuts $[0, 1]$ and $[m_3, m_2]$ on the corresponding Riemann surfaces $\Re_1$ and $\Re_2$ defined by

$$
\Re_1 : W^2 = (1 - z_1)z_1, \quad \Re_2 : W^2 = M(z_2). \tag{2.2.10}
$$

The pair (α_1, α_2) can be considered as an Abel–Jacobi map with singularities. These singularities result in an additional monodromy in the space of parameters.

Equations (2.2.5) define Lagrangian submanifolds $\mathcal{L} = \Re_1 \times \Re_2$ as level sets of the first integrals in the phase space $\mathbb{C}^4$ except at singular points, namely at points where z_1 is 0 or 1 and z_2 is ± 1.

Definition 2.1 *We call the Hamiltonian system with Hamiltonian* (2.2.4) *and first integrals* (2.2.5) *on the Lagrangian submanifolds* $\mathcal{L}$ *the complex Hamiltonian system associated with the* 2*-dimensional spherical pendulum.*

This family of parameterized Lagrangian submanifolds will be used to define semiclassical monodromy and Maslov phases below.

The second expression in (2.2.9) can be considered as a Jacobi inversion problem on the Riemann surface (2.2.8). This shows that z_2 is a periodic function defined on this Riemann surface. The first integral from the expression for α_1 is equal to θ_1. This shows that

$$\begin{aligned} \theta_1 &= \frac{1}{2}\int_{z_2^0}^{z_2} \frac{\beta_1 dz_2}{(1-z_2^2)\sqrt{M(z_2)}} - \alpha_1^0 \\ \theta_2 &= \arccos z_2. \end{aligned} \tag{2.2.11}$$

Here the function z_2, the upper limit of the integral in the expression for θ_1, is considered on the Riemann surface $\Re$ given by (2.2.8). Therefore, θ_1 is defined on the covering space of $\Re$. In the real case, these two angles (θ_1, θ_2) are the usual spherical coordinates of the pendulum.

In conclusion, note especially that the construction presented here resolves the problem of multivaluedness for the angle representations for the two dimensional spherical pendulum only for the fixed values of parameters β_1 and β_2. First, averaged values of α_1 and α_2 are multivalued functions of β_1 and β_2 due to presence of classical monodromy. (For details concerning classical monodromy see Duistermaat [1980], Cushman and Duistermaat [1988] and Bates and Zou [1993]). Second, the additional monodromy of the singularities in the space of parameters is related to the symplectic representation of the braid group in a way described in Alber [1991b] and Alber and Marsden [1992]. These two types of monodromy are obtained when the system goes through the singularities ($\beta_1 = 0, \beta_2 = \pm 1$) in the first case and $m_3 = m_2$ in the second case. The first type of singularity can be resolved in the real context. In the second case, complexification is essential. We notice that in the multidimensional case $n \geq 3$, one gets additional monodromy due to the presence of additional singularities.

3 The Homoclinic Variety for the Spherical Pendulum

In this section, we will investigate the dependence of the Hamiltonian system defined by the spherical pendulum, on parameters. We will choose the parameters, as in the previous section, to be the β_j introduced in the Appendix for the n dimensional case. In the two dimensional case, the parameters, β_1 and β_2 are the angular momentum and the energy, as in Cushman and Duistermaat [1988], as we mentioned earlier.

This will lead, in particular, to the description of complex exponential Hamiltonians. We will also construct explicit homoclinic action-angle variables and describe the corresponding monodromy. In the next section, this monodromy will be shown to be one of the components of the semiclassical monodromy. This is done by investigating the dependence of the complex modes and the quantum conditions on the parameters of the system.

3.1 Homoclinic Angle Representations We now will choose special values of β_1 and β_2 that correspond to choosing special values of the constants of motion that put one on the homoclinic variety. This yields the following form of the basic polynomial associated with the homoclinic variety

$$M(z) = (1-z)^2(1+z). \tag{3.3.1}$$

Namely, we consider the following limiting process: in the polynomial (2.2.8), we let β_1^2 tend to zero and let β_2^2 tend to one so that the polynomial develops a double root. Recall that this corresponds to a special value of the conserved quantities. In this case, the system of angle variables can be represented as follows,

$$\begin{aligned} \alpha_1 &= -\int_{z_1^0}^{z_1} \frac{1}{\sqrt{(1-z_1)z_1}}\, dz_1 = \alpha_1^0, \\ \alpha_2 &= -\sqrt{2}\int_{z_2^0}^{z_2} \frac{dz_2}{(1-z_2)\sqrt{1+z_2}} = t + \alpha_2^0. \end{aligned} \tag{3.3.2}$$

The problem of inversion for (3.3.2) leads to the expression for the homoclinic orbit of the usual pendulum. This means that homoclinic manifold for the spherical pendulum splits into 1-dimensional homoclinic orbits on the great circles obtained by intersecting S^2 and a plane through the poles parameterized by the angle $\theta_1 = \alpha_1^0 = \text{const}$.

Lastly, we construct exponential Hamiltonians on the homoclinic varieties.

Theorem 3.1 *The spherical pendulum on the homoclinic variety defined above is a Hamiltonian system with a Hamiltonian of exponential type.*

Proof We introduce the following action function

$$S = \int_{z_1^0}^{z_1} \sqrt{(1-z_1)z_1 + b_1}\, dz_1 + \int_{z_2^0}^{z_2} \frac{\log(b_2 - z_2)}{\sqrt{1+z_2}}\, dz_2. \tag{3.3.3}$$

This corresponds to a new choice of the first integrals

$$\begin{aligned} P_1 &= \sqrt{(1-z_1)z_1 + b_1}, \\ P_2 &= \frac{\log(b_2 - z_2)}{\sqrt{1+z_2}}, \end{aligned} \tag{3.3.4}$$

for the system of differential equations

$$\begin{aligned} z_1' &= \sqrt{(1-z_1)z_1 + b_1} \\ z_2' &= (b_2 - z_2)\sqrt{1+z_2}. \end{aligned} \tag{3.3.5}$$

This is obtained from the Hamiltonian system with Hamiltonian (5.5.1) by setting $\beta_1 = 0$ and $\beta_2 = 1$ and by introducing parameters b_1 and b_2. System (3.3.5) is a Hamiltonian system with the Hamiltonian

$$H = \left(P_1^2 - ((1-z_1)\, z_1 + b_1)\right) + \left(e^{\sqrt{1+z_2}P_2} - (b_2 - z_2)\right). \tag{3.3.6}$$

Now we consider the following system of action-angle variables

$$\begin{aligned} I_1 &= b_1, \quad \alpha_1 = -\frac{\partial S}{\partial b_1}, \\ I_2 &= b_2, \quad \alpha_2 = -\frac{\partial S}{\partial b_2}. \end{aligned} \tag{3.3.7}$$

Lastly, one obtains angle representation (3.3.2) by setting $b_1 = 0$ and $b_2 = 1$ in (3.3.7). □

4 Semiclassical Monodromy

In what follows, we recall from Alber [1989, 1991] and Alber and Marsden [1992], a method of complex geometric asymptotics for integrable Hamiltonian flows on Riemann surfaces. We will use geometric asymptotics to describe the quantization conditions of Bohr-Sommerfeld-Keller (BSK) type. Then we will investigate the dependence of these conditions on the parameters (i.e., the first integrals) of the system. This dependence near singularities produces effects caused by the classical and semiclassical monodromy.

Let us consider quadratic complex Hamiltonians of the following form:

$$H = \frac{1}{2}\sum_{j=1}^{n} g^{jj} P_{z_j}^2 + V(z_1, \dots, z_n) \tag{4.4.1}$$

defined on $\mathbb{C}^{2n}$. We think of $\mathbb{C}^{2n}$ as being the cotangent bundle of $\mathbb{C}^n$, with configuration variables $\mu_1, \dots, \mu_n$ and with canonically conjugate momenta $P_1, \dots, P_n$.

Notice that the systems considered earlier are indeed of this form. We consider the functions g^{jj} as components of a (diagonal) Riemannian metric, construct the associated Laplace-Beltrami operator, and then the stationary Schrödinger equation

$$\nabla^j \nabla_j U + w^2(E - V)U = 0, \tag{4.4.2}$$

defined on the n-dimensional complex Riemannian manifold $\mathbb{C}^n$. Here ∇^j and ∇_j are covariant and contravariant derivatives defined by the tensor g^{jj} and w (which is the inverse of Planck's constant $\hbar$) and E (the energy eigenvalue) are parameters. Note also that in general, the metric tensor is not constant, and even may have singularities, so that the kinetic term in the expression for H is *not* purely quadratic.

Now we establish a link between equation (4.4.2) and the Hamiltonian system (4.4.1) by means of geometric asymptotics; namely, we consider the following function that is similar to the well known Ansatz from WKB theory:

$$\begin{aligned} U(z_1, \dots, z_n) &= \sum_k A_k(z_1, \dots, z_n) \exp[iwS_k(z_1, \dots, z_n)] \\ &= \sum_k \prod_{j=1}^{n} U_{kj}(z_j) \\ &= \sum_k \prod_{j=1}^{n} (A_{kj}(z_j) \exp[iwS_{kj}(z_j)]), \end{aligned} \tag{4.4.3}$$

which is a multivalued function of several complex variables defined on $\mathbb{C}^n$. If, instead, one considers U to be defined on the covering space of the Jacobi variety of the problem, then U becomes single valued. (The Jacobi variety was described in Section 3.) The functions present in this expression together with r, which denotes a vector of Maslov indices, will be determined below.

Substituting (4.4.3) in (4.4.2), equating coefficients for w and w^2 and integrating, we obtain the amplitude function A, which is a solution of the transport equation, of the form

$$A = \frac{A_0}{\sqrt{(D \det J)}}. \tag{4.4.4}$$

Here D is the volume element of the metric: $D = \sqrt{\prod_{l=1}^{n} g_{ll}}$ and J is the Jacobian of the change of coordinates from the z-representation to the angle (α) representation. We also find that the phase function S is a solution of the Hamilton-Jacobi equation

$$\Delta^j S \Delta_j S - V = E, \tag{4.4.5}$$

meaning that it coincides with the action function.

Now we can apply the above construction to the 2-dimensional spherical pendulum. Note that in this case the action function (2.2.6) can be represented in terms of angle variables (2.2.9) as follows

$$S = -\beta_1 \alpha_1^0 - \beta_2^2 \alpha_2 - 2 \int_{z_2^0}^{z_2} \frac{z_2 dz_2}{\sqrt{M(z_2)}}. \tag{4.4.6}$$

The last two terms correspond to the holomorphic and meromorphic parts of the action function. The holomorphic part is proportional to the angle variable of the classical problem. The amplitude A can be found after calculating D and J

$$D = \sqrt{g_{11} g_{22}} = \frac{1}{\sqrt{(1 - z_1) z_1}} \tag{4.4.7}$$

and

$$\det J^{-1} = \left| \frac{\partial \alpha_i}{\partial z_j} \right| = -\frac{2\beta_2}{\sqrt{M(z_2) z_1 (1 - z_1)}}. \tag{4.4.8}$$

This results in the following form of the function U:

$$U = \sum_{k=(k_1,k_2)} A_0 \sqrt{2\beta_2} (M(z_2))^{-\frac{1}{4}} \exp \left[iw \sum_{j=1}^{2} S_{kj}(z_j) \right], \tag{4.4.9}$$

where

$$S_{k1}(z_1) = \int_{z_1^0}^{z_1} \sqrt{\frac{\beta_1^2}{(1 - z_1) z_1}}\, dz_1 + k_1 T_1, \tag{4.4.10}$$

and

$$S_{k2}(z_2) = \int_{z_2^0}^{z_2} \sqrt{\frac{M(z_2)}{(1 - z_2^2)^2}}\, dz_2 + k_2 T_2 + \frac{\pi}{2} r_2, \tag{4.4.11}$$

where r_2 is Maslov index, and

$$\begin{aligned} T_1 &= \oint_{l_1} \sqrt{\frac{\beta_1^2}{(1 - z_1) z_1}}\, dz_1, \\ T_2 &= 2\beta_2^2 \oint_{l_2} \frac{dz_2}{\sqrt{M(z_2)}} - 2 \oint_{l_2} \frac{z_2 dz_2}{\sqrt{M(z_2)}} - \beta_1^2 \oint_{l_2} \frac{dz_2}{(1 - z_2^2)\sqrt{M(z_2)}}. \end{aligned} \tag{4.4.12}$$

The amplitude A has singularities at the branch points $z_2 = m_1, m_2, m_3$ of the Riemann surface (2.2.8). Each time a trajectory approaches one of these singularities, we continue in complex time and go around a small circle in complex plain enclosing the singularity. This results in a phase shift ($\pm i\pi/2$) of the phase function S, which is common in geometric asymptotics. The indices k_1 and k_2 keep track of the number of oriented circuits for z_1 and z_2 around l_1 and l_2. The complex mode (4.4.9) is defined on the covering space of the complex Jacobi variety. Note that in the real case, it is defined on the covering space of a real subtorus. Keeping this in mind, quantum conditions of BSK type can be imposed as conditions on the number of sheets of the covering space of the corresponding Riemann surface for each coordinate z_j:

$$\begin{aligned} wk_1T_1 &= 2\pi N_1, \\ \frac{\pi}{2}r_2 + wk_2T_2 &= 2\pi N_2. \end{aligned} \tag{4.4.13}$$

Here N_1, N_2 are integer quantum numbers. The quantum conditions (4.4.13) include a monodromy part after transport along a closed loop in the space of parameters (β_1 and β_2). This semiclassical monodromy consists of a classical part as well as a contribution from complex monodromy and the Maslov phase.

Classical monodromy may be explained briefly as follows. We consider two different cases, namely the case $-1 < \beta_2^2 < 1$ and $\beta_2^2 > 1$. In the first case, one considers a cycle l_2 over the cut $[-1, \beta_2^2]$ and in the second case, one considers a cycle l_2 over the cut $[-1, 1]$. There is a closed curve in the space of parameters that leads one from one case to the other. Evidently, there is a difference in the values of the third integral in the expression for T_2 between the two cases that is given by the residue of the integrand at $z_2 = 1$.

The complex monodromy can be demonstrated if one of the roots m_1 and m_2 of the basic polynomial $M(z)$ approach each other. This singularity can be resolved by interchanging these two roots in the complex plane so as to avoid a real singularity. This leads to the change of orientation of the cycle l_2 and in the general case, can be described by the generator of the symplectic representation of the braid group. It results in additional shift in the quantum conditions.

The third type of shift in the quantum conditions comes from the integral representation for the Maslov class.

The complex mode (4.4.9), which corresponds to a particular choice of parameters in (4.4.13) is similar to an acoustic mode that occurs in the whispering gallery phenomenon described in Keller and Rubinow [1960] and Alber [1989, 1991].

The n-dimensional system can be treated in a similar way. A complex mode U has the form

$$U = \sum_{k=(k_1,\dots,k_n)} \frac{A_0\sqrt{2\beta_2\dots\beta_n}}{((-1)^n K_2(z_2)\dots K_n(z_n)(1-z_3)(1-z_4)^2\dots(1-z_n)^{n-2})^{\frac{1}{4}}} \times \exp\left[iw\sum_{j=1}^{n} S_{kj}(z_j)\right]. \tag{4.4.14}$$

which yields following quantum conditions

$$\begin{aligned} wk_1T_1 &= 2\pi N_1, \\ \frac{\pi}{2}r_2 + wk_2T_2 &= 2\pi N_2, \\ &\dots \\ \frac{\pi}{2}r_n + wk_nT_n &= 2\pi N_n. \end{aligned} \tag{4.4.15}$$

Since the form of $T_j, j = 2, \dots, n-1$ is different from both T_1 and T_n, one gets additional new types of monodromy in the n-dimensional case.

Lastly, the results of Section 4 yield a construction of the complex mode defined on the homoclinic varieties despite the fact that the Hamiltonians are not quadratic, but are exponential in this case.

5 Appendix: The n-dimensional Spherical Pendulum

Recall that a Riemann surface $\mathfrak{R}$ is a connected two-dimensional topological manifold with a complex-analytic structure. Compact Riemann surfaces are determined by nonsingular algebraic curves of the form $R(W, E) = 0$, where (W, E) are points in $\mathbb{C}^2$, and $R(W, E)$ is a polynomial with no multiple roots. We shall also need notions about complex Jacobi varieties and Abel–Jacobi maps; for details see Ercolani and McKean [1990].

A.1 The Hamiltonian and First Integrals In what follows we will handle the spherical pendulum in a way similar to the problem of geodesics on an n-dimensional sphere in $\mathbb{R}^{n+1}$.

We recall from Alber and Alber [1985] the following form of the Hamiltonian

$$H = \frac{1}{2} \sum_{j=1}^{n} P_{z_j}^2 (1 - z_j) z_j \left(\prod_{k=j+1}^{n} \frac{1}{1 - z_k} \right). \tag{5.5.1}$$

for the problem of geodesics on the n-sphere of radius 1:

$$\sum_{j=1}^{n+1} Q_j^2 = 1.$$

To explain the notation in this expression for the Hamiltonian, we consider n-dimensional spherical coordinates on the unit n-sphere defined as follows

$$\begin{aligned} Q_1 &= (\cos\theta_1)(\sin\theta_2)\dots(\sin\theta_n), \\ Q_2 &= (\sin\theta_1)(\sin\theta_2)\dots(\sin\theta_n), \\ &\dots \\ Q_n &= (\cos\theta_{n-1})(\sin\theta_n), \\ Q_{n+1} &= \cos\theta_n. \end{aligned} \tag{5.5.2}$$

In (5.5.1) the variables z_j and P_{z_j}, the momentum conjugate to z_j, are related to spherical coordinates and momentum (θ_j, P_{θ_j}) on the sphere by

$$\begin{aligned} z_j &= \cos^2\theta_j, \\ P_{z_j}^2 (1 - z_j) z_j &= P_{\theta_j}^2, \end{aligned} \tag{5.5.3}$$

where $j = 1, \dots, n$. The Hamiltonian is taken to be the standard one, namely the kinetic energy. When the Hamiltonian is transformed to the variables z_j and P_{z_j} one gets the expression (5.5.1). Spherical coordinates provide a convenient "nested" structure that leads to the first integrals, whereas the variables z_j provide variables on the Riemann surfaces determined by the first integrals. We will demonstrate this below explicitly for the spherical pendulum.

The Hamiltonian of the n-dimensional spherical pendulum in Cartesian coordinates Q_j and their conjugate momenta P_j has the form

$$H = \frac{1}{2}\left(\sum_{j=1}^{n+1} P_j^2 - \left(\sum_{j=1}^{n+1} P_j Q_j\right)^2\right) + Q_{n+1}. \tag{5.5.4}$$

Here the acceleration due to gravity and constraint of the length of Q have been set equal to one.

Remark 5.1 In this Hamiltonian the constraint

$$\sum_{j=1}^{n+1} Q_j^2 = 1 \tag{5.5.5}$$

has been enforced by the extra term

$$\left(\sum_{j=1}^{n+1} P_j Q_j\right)^2.$$

Hamilton's equations on $\mathbb{R}^{2(n+1)}$ for (5.5.4) automatically preserve the set (5.5.5). Notice also that this extra term vanishes on the constrained phase space, so the Hamiltonian becomes the usual expression of kinetic energy plus potential energy on this constrained phase space. This construction provides an example of a Hamiltonian that automatically preserves a desired set of constraints.

The same Hamiltonian in the n-dimensional spherical coordinates (5.5.2) can be expressed in the following "nested" form

$$H = \frac{1}{2R^2}\sum_{j=1}^{n} P_{\theta_j}^2 \left(\prod_{k=j+1}^{n} \frac{1}{(\sin\theta_k)^2}\right) + R\cos\theta_n, \tag{5.5.6}$$

or

$$\begin{aligned} H = \frac{1}{2R^2}\left(\frac{1}{\sin^2\theta_n}\left(P_{\theta_{n-1}}^2 + \frac{1}{\sin^2\theta_{n-1}}\left(P_{\theta_{n-2}}^2 + \frac{1}{\sin^2\theta_{n-2}}\right.\right.\right. \\ \left.\left.\left.\left(\cdots + \frac{1}{\sin^2\theta_3}\left(P_{\theta_2}^2 + \frac{1}{\sin^2\theta_2}P_{\theta_1}^2\right)\right)\cdots\right)\right)\right) \\ + \frac{1}{2R^2}P_{\theta_n}^2 + R\cos\theta_n. \end{aligned} \tag{5.5.7}$$

The change of coordinates

$$\begin{aligned} z_j &= \cos^2\theta_j, \quad P_{z_j}^2(1-z_j)z_j = P_{\theta_j}^2, \qquad j = 1,\ldots,(n-1); \\ z_n &= \cos\theta_n, \qquad P_{z_n}^2(1-z_n^2) = P_{\theta_n}^2. \end{aligned} \tag{5.5.8}$$

results in the Hamiltonian

$$H = \frac{1}{2}\sum_{j=1}^{n-1} P_{z_j}^2 (1 - z_j) z_j \left(\prod_{k=j+1}^{n-1} \frac{1}{(1 - z_k)} \right) \frac{1}{(1 - z_n^2)} + \frac{1}{2} P_{z_n}^2 (1 - z_n^2) + z_n. \quad (5.5.9)$$

Here $R = 1$ and (z_j, P_{z_j}) and (θ_j, P_{θ_j}) are pairs of conjugate variables.

The nested structure of the Hamiltonian (5.5.7) shows that one has the following first integrals for the n-dimensional pendulum

$$\begin{aligned} P_{\theta_1}^2 &= \beta_1^2, \\ P_{\theta_2}^2 + \frac{\beta_1^2}{\sin^2 \theta_2} &= \beta_2^2, \\ &\dots \\ P_{\theta_{n-1}}^2 + \frac{\beta_{n-2}^2}{\sin^2 \theta_{n-1}} &= \beta_{n-1}^2, \\ \frac{1}{2}\left(P_{\theta_n}^2 + \frac{\beta_{n-1}^2}{\sin^2 \theta_n} \right) + \cos\theta_n &= \beta_n^2. \end{aligned} \quad (5.5.10)$$

Here β_j are constants along solutions of the corresponding Hamiltonian system. Denote

$$K_j(z) = \beta_j^2 (1 - z) - \beta_{j-1}^2, \quad j = 2, \dots, n-1,$$

and

$$K_n(z) = 2\left(\beta_n^2 - z\right)\left(1 - z^2\right) - \beta_{n-1}^2.$$

Using the change of coordinates (5.5.8), the first integrals take the form

$$\begin{aligned} P_{z_1}^2 &= \frac{\beta_1^2}{z_1(1 - z_1)}, \\ P_{z_2}^2 &= \frac{\beta_2^2}{z_2(1 - z_2)} - \frac{\beta_1^2}{z_2(1 - z_2)^2} = \frac{K_2(z_2)}{z_2(1 - z_2)^2}, \\ &\dots \\ P_{z_{n-1}}^2 &= \frac{\beta_{n-1}^2}{z_{n-1}(1 - z_{n-1})} - \frac{\beta_{n-2}^2}{z_{n-1}(1 - z_{n-1})^2} = \frac{K_{n-1}(z_{n-1})}{z_{n-1}(1 - z_{n-1})^2}, \\ P_{z_n}^2 &= 2\frac{\beta_n^2}{(1 - z_n^2)} - \frac{\beta_{n-1}^2}{(1 - z_n^2)^2} - 2\frac{z_n}{(1 - z_n^2)} = \frac{K_n(z_n)}{(1 - z_n^2)^2}, \end{aligned} \quad (5.5.11)$$

where constants β_j^2 are positive real numbers. Therefore, we consider the real-valued solutions in which case one has a correspondence with the system in spherical coordinates.

A.2 The Complexified n-dimensional Spherical Pendulum Now we use the representation (5.5.11) to extend our system into the complex domain by considering β_j^2 as complex numbers and variables z_j defined on the associated Riemann surfaces:

$$
\begin{aligned}
\Re_1 : W_1^2 &= \frac{\beta_1^2}{z_1(1-z_1)}, \\
\Re_2 : W_2^2 &= \frac{K_2(z_2)}{z_2(1-z_2)^2}, \\
&\cdots \\
\Re_{n-1} : W_{n-1}^2 &= \frac{K_{n-1}(z_{n-1})}{z_{n-1}(1-z_{n-1})^2} \\
\Re_n : W_n^2 &= \frac{K_n(z_n)}{(1-z_n^2)^2}.
\end{aligned}
\tag{5.5.12}
$$

Equations (5.5.10) define Lagrangian submanifolds $\mathcal{L} = \Re_1 \times \cdots \times \Re_n$ as level sets of the first integrals in the phase space $\mathbb{C}^{2n}$ except at singular points, namely at points where z_j is 0 or ± 1.

Definition 5.2 *We call the Hamiltonian system with Hamiltonian* (5.5.9) *and first integrals* (5.5.11) *on the Lagrangian submanifolds $\mathcal{L}$ the complex Hamiltonian system associated with the n-dimensional spherical pendulum.*

A.3 Monodromy in the n-dimensional Case We construct an action function of canonical transformation from (z, P_z) to the new (α, β) coordinates to be described below, in the form

$$
S = \sum_{j=1}^{n} \int_{z_j^0}^{z_j} P_{z_j} dz_j. \tag{5.5.13}
$$

Here expressions for P_{z_j} are taken from (5.5.11). Notice that the function S is, at the same time, a generating equation of the Lagrangian submanifold in the phase space.

We choose β_j and $\alpha_j = -\partial S/\partial\beta_j$ as conjugate action-angle variables in the n-dimensional case. Thus,

$$
\begin{aligned}
\alpha_1 &= -\int_{z_1^0}^{z_1} \frac{1}{\sqrt{(1-z_1)z_1}}\,dz_1 + \int_{z_2^0}^{z_2} \frac{\beta_1 dz_2}{(1-z_2)\sqrt{z_2 K_2(z_2)}}, \\
&\dots \\
\alpha_j &= -\int_{z_j^0}^{z_j} \frac{\beta_j dz_j}{\sqrt{z_j K_j(z_j)}} + \int_{z_{j+1}^0}^{z_{j+1}} \frac{\beta_j dz_{j+1}}{(1-z_{j+1})\sqrt{z_{j+1}K_{j+1}(z_{j+1})}}, \\
&\dots \\
\alpha_{n-1} &= -\int_{z_{n-1}^0}^{z_{n-1}} \frac{\beta_{n-1} dz_{n-1}}{\sqrt{z_{n-1}K_{n-1}(z_{n-1})}} + \int_{z_n^0}^{z_n} \frac{\beta_{n-1} dz_n}{(1-z_n^2)\sqrt{K_n(z_n)}}, \\
\alpha_n &= -\int_{z_n^0}^{z_n} \frac{2\beta_n dz_n}{\sqrt{K_n(z_n)}}.
\end{aligned}
\tag{5.5.14}
$$

where $j = 2, \dots, n-2$. This vector function $\alpha_1, \dots, \alpha_n$ defines the Abel–Jacobi map and its image (i.e., the space in which these angle variables are uniquely defined) is called the Jacobi variety of the problem.

Theorem 5.3 *The Hamiltonian flow of the n-dimensional spherical pendulum linearizes in terms of the action-angle variables* (5.5.14) *on each Jacobi variety determined by a choice of parameters $\beta_1, \dots, \beta_n$.*

Proof We substitute in each equation of a Hamiltonian system $z_j' = \partial H/\partial P_{z_j}$ with the Hamiltonian (5.5.9) expression for momenta P_{z_j} from (5.5.11) to obtain the following system of differential equations on the corresponding Riemann surfaces

$$
\begin{aligned}
z_1' &= \sqrt{(1-z_1)z_1}\left(\prod_{k=2}^{n-1}\frac{1}{(1-z_k)}\right)\frac{1}{(1-z_n^2)}, \\
&\dots \\
z_j' &= \sqrt{K_j(z_j)z_j}\left(\prod_{k=j+1}^{n-1}\frac{1}{(1-z_k)}\right)\frac{1}{(1-z_n^2)}, \\
&\dots \\
z_n' &= \sqrt{K_n(z_n)},
\end{aligned}
\tag{5.5.15}
$$

where $j = 1, \dots, n-1$. Transforming equations (5.5.15) and taking linear combination one obtains

$$\begin{aligned}
\frac{\beta_1 z'_2}{(1-z_2)\sqrt{z_2 K_2(z_2)}} - \frac{z'_1}{\sqrt{(1-z_1)z_1}} &= 0, \\
&\dots \\
\frac{\beta_j z'_{j+1}}{(1-z_{j+1})\sqrt{z_{j+1} K_{j+1}(z_{j+1})}} - \frac{\beta_j z'_j}{\sqrt{z_j K_j(z_j)}} &= 0, \\
&\dots \\
\frac{\beta_{n-1} z'_n}{(1-z_n^2)\sqrt{K_n(z_n)}} - \frac{\beta_{n-1} z'_{n-1}}{\sqrt{z_{n-1} K_{n-1}(z_{n-1})}} &= 0, \\
- \frac{2\beta_n z'_n}{\sqrt{K_n(z_n)}} &= \beta_n.
\end{aligned} \tag{5.5.16}$$

where $\beta_1, \dots, \beta_n$ are constants and $j = 2, \dots, n-2$.

After integrating (5.5.16) we obtain on the left-hand side expressions which coincide with the expressions (5.5.14) for the angle variables

$$\begin{aligned}
\alpha_j &= \alpha_j^0, \\
\alpha_n &= \beta_n t + \alpha_n^0.
\end{aligned} \tag{5.5.17}$$

where $j = 1, \dots, n-1$. □

A.4 Homoclinic Varieties in the n-dimensional case We fix an integer m and apply the following limiting processes to (5.5.14) in the given order:

$$\begin{aligned}
\beta_1 &\to 0, \\
&\dots \\
\beta_m &\to 0.
\end{aligned} \tag{5.5.18}$$

This yields the following expression for the angle representation

$$
\begin{aligned}
\alpha_1 &= -\int_{z_1^0}^{z_1} \frac{dz_1}{\sqrt{(1-z_1)z_1}} = \alpha_1^0, \\
&\dots \\
\alpha_m &= -\int_{z_m^0}^{z_m} \frac{dz_m}{\sqrt{(1-z_m)z_m}} = \alpha_m^0, \\
\alpha_{m+1} &= -\int_{z_{m+1}^0}^{z_{m+1}} \frac{dz_{m+1}}{\sqrt{z_{m+1}(1-z_{m+1})}} \\
&\quad + \int_{z_{m+2}^0}^{z_{m+2}} \frac{\beta_{m+1} dz_{m+2}}{(1-z_{m+2})\sqrt{z_{m+2}K_{m+2}(z_{m+2})}}, \\
&\dots \\
\alpha_j &= -\int_{z_j^0}^{z_j} \frac{\beta_j dz_j}{\sqrt{z_j K_j(z_j)}} \\
&\quad + \int_{z_{j+1}^0}^{z_{j+1}} \frac{\beta_j dz_{j+1}}{(1-z_{j+1})\sqrt{z_{j+1}K_{j+1}(z_{j+1})}} = \alpha_j^0, \\
&\dots \\
\alpha_{n-1} &= -\int_{z_{n-1}^0}^{z_{n-1}} \frac{\beta_{n-1} dz_{n-1}}{\sqrt{z_{n-1}K_{n-1}(z_{n-1})}} \\
&\quad + \int_{z_n^0}^{z_n} \frac{\beta_{n-1} dz_n}{(1-z_n^2)\sqrt{K_n(z_n)}} = \alpha_{n-1}^0, \\
\alpha_n &= -\int_{z_n^0}^{z_n} \frac{2\beta_n dz_n}{\sqrt{K_n(z_n)}} = \beta_n t + \alpha_n^0,
\end{aligned}
\tag{5.5.19}
$$

where $j = m+2, \dots, n-2$.

There are two different cases. For $m < n-1$, the system (5.5.19) describes an m-dimensional family of parameterized $(n-m)$-dimensional spherical pendula. For $m = n-1$ and $\beta_n = 1$, one obtains a family of 1-dimensional homoclinic orbits.

There have been many important developments in which the methods of complex and algebraic geometry have been used to investigate the eigenfunctions of Hill's operator in the context of integrable equations. In Alber and Marsden [1995] we link a new class of Hamiltonian systems on Riemann surfaces to systems of pde's using Bloch eigenfunctions for stationary Shrödinger equations with new types of potentials. In particular, this yields a system of pde's with monodromy.

Acknowledgments We thank Richard Cushman for a very stimulating discussion on monodromy for the spherical pendulum and for the suggestion that we apply our techniques to it. Mark Alber thanks the Institute for Advanced Study in Princeton for the hospitality during the fall of 1993, and the Mathematical Sciences Research Institute, Berkeley during the spring of 1994.

This research was partially supported by NSF grants DMS 9022140 and 9403861 (M. S. Alber), and NSF grant DMS 9302992 (J.E. Marsden).

References

Ablowitz, M. J. and Segur, H. [1981], *Solitons and the Inverse Scattering Transform*, SIAM, Philadelphia.

Abraham, R. and Marsden, J. E. [1978], *Foundations of Mechanics*, New York–Amsterdam.

Adler, M. and Van Moerbeke, P. [1980], *Completely integrable systems, Kac-Moody Lie algebras and curves*, Adv. in Math., **38**, 267-317.

Alber, M. S. [1992], *Complex Geometric Asymptotics, Geometric Phases and Nonlinear Integrable Systems*, Studies in Math. Phys. **3**, North-Holland, Elsevier Science Publishers B.V., Amsterdam, 415–427.

Alber, M. S. [1991a], *Hyperbolic Geometric Asymptotics*, Asymptotic Analysis, **5**, 161–172.

Alber, M. S. [1991b], *Geometric Phases and Braid Groups*, (S. Catto and A. Rocha ed.), Proc. of the XXTH International Conference on Differential Geometric Methods, The City University of New York, World Scientific.

Alber, M. S. [1989], *On integrable systems and semiclassical solutions of the stationary Schrödinger equations*, Inverse Problems, **5**, 131–148.

Alber S. J. [1991], *Associated integrable systems*, J. Math. Phys., **32**, 916–922.

Alber, M. S. and Alber, S. J. [1985], *Hamiltonian formalism for finite-zone solutions of integrable equations*, C.R. Acad. Sc. Paris, **301**, 777–781.

Alber, M. S., Camassa, R., Holm, D. D., and Marsden, J. E. [1994], *The geometry of peaked solitons and billiard solutions of a class of integrable pde's*, Lett. Math. Phys., **32**, 137–151.

Alber, M. S., Camassa, R., Holm, D. D., and Marsden, J. E. [1995], *On the link between umbilic geodesics and soliton solutions of nonlinear PDE's*, Proc. Roy. Soc. A, to appear.

Alber, M. S. and Marsden J. E. [1992], *On Geometric Phases for Soliton Equations*, Commun. Math. Phys., **149**, 217–240.

Alber, M. S. and Marsden J. E. [1994a], *Geometric phases and monodromy at singularities*, Singular Limits of Dispersive Waves, NATO ASI Series C, **320**, (N. Ercolani et al, eds.) Plenum Press, NY, 273–295.

Alber, M. S. and Marsden, J. E. [1994b], *Resonant geometric phases for soliton equations*, The Fields Institute Commun., **3**, 1–26.

Alber, M. S. and Marsden, J. E. [1955], *Complex WKB Solutions and Hamiltonian Systems on Riemann Surfaces*, Hewlett Packard Research Lab. Preprint HPL-BRIMS-95-002.

Arnold, V. I. [1967], *Characteristic class entering in quantization conditions*, Funct. Anal. Appl., **1**, 1–8.

Arnold, V. I. [1972], *Modes and quasimodes*, Funct. Anal. Appl. **6**, 12–20.

Baider, A., Churchill, R. C., and Rod, D. L. [1990], *Monodromy and Nonintegrability in Complex Hamiltonian Systems*, J. Dynamics and Diff. Equations, **4**, 451.

Bates, L. and Zou, M. [1993], *Degeneration of Hamiltonian monodromy cycles*, Nonlinearity, **6**, 313–335.

Creagh, S. C., Robbins, J. M., and Littlejohn, R. G. [1990], *Geometric properties of Maslov indices in the semiclassical trace formula for density of states*, Phys. Rev. A, **42**, 1907.

Cushman, R. and Duistermaat, J. J. [1988], *The Quantum Spherical Pendulum*, Bull.Amer.Math.Soc. (new series), **19**, 475–479.

Duistermaat, J. J. [1980], *On global Action-Angle Coordinates*, Comm. Pure Appl. Math., **23**, 687-706.

Duistermaat, J. J. [1988], *The Quantum Spherical Pendulum*, (M. Cahen and M.Flato eds.), Quantum Theories and Geometry, Kluwer Academic Publiishers, 53–64.

Ercolani, N. [1989], *Generalized Theta functions and homoclinic varieties*, Proc. Symp. Pure Appl. Math., **49**, 87–100.

Ercolani, N. and McKean, H. P. [1990], *Geometry of KdV(4): Abel sums, Jacobi variety, and theta function in the scattering case*, Invent. Math., **99**, 483.

Guillemin, V. and Sternberg, S. [1977], *Geometric Asymptotics,* Math.Surveys, **14**, AMS Providence, Rhode Island.

Guillemin, V. and Sternberg, S. [1983], *The Gelfand-Cetlin system and the quantization of complex Flag Manifolds*, J. Funct. Anal., **52**, 106–128.

Guillemin, V. and Uribe, A. [1989], *Monodromy in the Quantum Spherical Pendulum*, Commun. Math. Phys., **122**, 563–574.

Horosov, E. [1990], *Perturbation of the spherical pendulum and Abelian integrals*, J. reine angew. Math., **408**, 114–135.

Keller, J. B. and Rubinow, S. I. [1960], *Asymptotic solution of eigenvalue problems*, Ann. Phys., **9**, 24.

Robbins, J. M. [1991], *Maslov indices in the Gutzwiller trace formula*, Nonlinearity, **4**, 343–363.

Fields Institute Communications
Volume 8, 1996

Hamiltonian Structure of Plane-wave Instabilities

Thomas J. Bridges
Department of Mathematical and Computing Sciences
University of Surrey
Guildford GU2 5XH, England

1 Introduction

One of the basic solutions of a translation-invariant Hamiltonian evolution equation, say the wave equation or its nonlinear counterparts, is the plane wave solution. For a linear equation, in one space dimension plus time, we might write such a wave as the real part of $Ae^{i(kx-\omega t)}$, $A \in \mathbb{C}$, where k is a wavenumber and ω a frequency. An entirely elementary observation is that such a wave is periodic in x of wavelength $\lambda = 2\pi/k$ and periodic in t of period $T = 2\pi/\omega$ and the two periodic directions can be identified with the coordinates of a two-torus. Given the presence of a Hamiltonian structure we might think of identifying the "plane-wave two-torus" with an invariant toroid of a 2-DOF classical Hamiltonian system, or at least have some features in common. The purpose of the present paper is to study such a structure for nonlinear plane waves. To determine this structure the governing equations for the nonlinear problem are written in the form

$$\mathbf{M}Z_t + \mathbf{K}Z_x = dS(Z) \tag{1.1}$$

where $\mathbf{M}$ and $\mathbf{K}$ are skew-symmetric operators and Z is some suitable set of coordinates. The functional $S(Z)$ is the (space-time) Hamiltonian function for the system and differs from the Hamiltonian function for the evolution equation. In other words we normally think of Hamiltonian evolution equations in the form

$$U_t = \mathcal{J}(U) \cdot dH(U)$$

with $\mathcal{J}(U)$ a skew-symmetric operator and $H(U)$ the Hamiltonian function in the usual sense. Rewriting (1.1) as $\mathbf{M}Z_t = dS(Z) - \mathbf{K}Z_x$ we might think of $dS - \mathbf{K}Z_x$ as a decomposition of the usual Hamiltonian function into two parts. In essence new coordinates are introduced into the evolution equation so that the new Hamiltonian function S does not contain any space derivatives.

It is the form (1.1) of the equations that evokes the toral structure of plane waves. This can be seen as follows. The terms $\mathbf{M}Z_t$ and $\mathbf{K}Z_x$ are gradients of functionals that play the same role that action functionals play in finite-dimensional

1991 *Mathematics Subject Classification.* 76D33, 35Q55, 58F05.

Hamiltonian systems. To highlight this relation define the (time)-action A and the (space)-action B so that

$$dA(Z) = \mathbf{M}Z_t \quad \text{and} \quad dB(Z) = \mathbf{K}Z_x\,.$$

Then (1.1) becomes $dS(Z) = dA(Z) + dB(Z)$. Now, suppose Z is restricted to functions that are periodic in x and t. Define new independent variables $\xi_1 = \omega t$ and $\xi_2 = kx$; then Z is 2π periodic in ξ_1 and ξ_2 and (1.1) transforms to

$$\omega\mathbf{M}\frac{\partial Z}{\partial \xi_1} + k\mathbf{K}\frac{\partial Z}{\partial \xi_2} = dS(Z) \tag{1.2}$$

or equivalently

$$\omega dA(Z) + k dB(Z) = dS(Z) \tag{1.3}$$

where A, B and S are now averaged over x and t. Equation (1.3) has the appearance of a variational principle of the following form: periodic plane waves are critical points of the Hamiltonian functional $S(Z)$ on level sets of the constraints $A(Z)$ and $B(Z)$ and the expression (1.3) is simply the necessary condition with ω and k as Lagrange multipliers. Before discussing this in more detail we record some facts about invariant toroids of classical finite-dimensional Hamiltonian systems.

There is a strong connection between (1.2)-(1.3) and variational principles for invariant toroids of 2-DOF Hamiltonian systems. Consider the 2-DOF Hamiltonian system

$$\mathbf{J}\Upsilon_t = dH(\Upsilon)\,, \quad \Upsilon = (q,p) \in \mathbb{R}^4\,, \quad \mathbf{J} = \begin{pmatrix} 0 & -I \\ I & 0 \end{pmatrix} \tag{1.4}$$

Upon restriction of Υ to an invariant toroid; letting $\xi_1 = \omega_1 t$ and $\xi_2 = \omega_2 t$ and considering $\Upsilon(\xi_1, \xi_2)$, the governing equation (1.4) is transformed to

$$\omega_1\mathbf{J}\frac{\partial \Upsilon}{\partial \xi_1} + \omega_2\mathbf{J}\frac{\partial \Upsilon}{\partial \xi_2} = dH(\Upsilon)\,. \tag{1.5}$$

Percival [1974] has given a variational principle for invariant toroids of fixed action that leads to the governing equation (1.5). It is constructed as follows. Define the averaging operator

$$\oint_{\mathbb{T}^2} F(\xi)\, d\xi = \frac{1}{(2\pi)^2}\int_0^{2\pi}\int_0^{2\pi} F(\xi)\, d\xi_1 d\xi_2 \tag{1.6}$$

and introduce independent action functions on the torus,

$$I_j = \oint_{\mathbb{T}^2}\left(p, \frac{\partial q}{\partial \xi_j}\right) d\xi \qquad j = 1, 2. \tag{1.7}$$

Then, with suitable averaging of H, solutions of (1.5) correspond to stationary points of the functional

$$\mathcal{F} = \oint_{\mathbb{T}^2} H(\Upsilon)\, d\xi - \omega_1 I_1 - \omega_2 I_2 \tag{1.8}$$

or equivalently this is a constrained variational problem with ω_1 and ω_2 as Lagrange multipliers (Percival [1974, p. 798]).

Comparison of (1.2) with (1.5) shows a remarkable similarity. The only essential difference is that the two skew-symmetric operators $\mathbf{M}$ and $\mathbf{K}$ in (1.5) differ and in general do not commute. However this fact is irrelevant to the variational principle (although it may be important in other contexts). This leads to the following variational principle for periodic plane waves of infinite-dimensional Hamiltonian systems.

Theorem 1.1 *(Variational principle for periodic plane waves). Let $S(Z)$, $A(Z)$ and $B(Z)$ be the S-Hamiltonian function, time-action and space-action respectively, for a Hamiltonian evolution equation on the real line, averaged over $\mathbb{T}^2$. Then periodic plane wave solutions correspond to critical points of S on level sets of A and B. Equivalently, with ω, the frequency, and k, the wavenumber, as Lagrange multipliers, periodic plane waves are critical points of the unconstrained functional*

$$\mathcal{F} = \oint_{\mathbb{T}^2} [S(Z) - \omega A(Z) - kB(Z)]\, d\xi. \tag{1.9}$$

Formally the proof follows from the arguments in (1.1)-(1.3); in particular applications the differentiablity of the functionals must be checked and so forth. Formal application of the result is straightforward and relies only on the ability to recast the Hamiltonian evolution equation $U_t = J\,dH$ in the form (1.1). In the examples in Sections 2-4 we will see that this is also straightforward.

The above variational principle is certainly not the only one for periodic plane waves (Bridges [1992b, Section 4] suggests two other variational principles). The energy-momentum method for example treats plane waves as critical points of the energy on level sets of the momentum. Or, with $c \in \mathbb{R}$, the wavespeed, as a Lagrange multiplier, critical points of the unconstrained functional $E - cI$ (with E the energy and I the momentum *integrated over a fixed wavelength*). The difficulty with the energy-momentum variational principle is that the wavelength is considered fixed and it is the ability to vary the wavelength that leads to plane-wave instability! So a (periodic) plane wave that minimizes the energy on level sets of the momentum may still be unstable (to perturbations of a different wavelength). The novelty of the space-time variational principle with functional (1.6) is that stability information can be extracted.

There are a number of observations that follow immediately from the variational principle for plane waves. To distinguish the value of a level set of the constraint set from its functional form define

$$I_1 = A(Z) \qquad \textit{and} \qquad I_2 = B(Z). \tag{1.10}$$

Then ω, k can be expressed in the following interesting form

$$\omega = \frac{\partial S}{\partial I_1} \qquad \textit{and} \qquad k = \frac{\partial S}{\partial I_2}. \tag{1.11}$$

To see this differentiate (1.7) with respect to I_1,

$$\frac{\partial A}{\partial I_1} = \langle dA(Z), Z_1\rangle = 1 \qquad \frac{\partial B}{\partial I_1} = \langle dB(Z), Z_1\rangle = 0$$

where $Z_1 = \partial Z/\partial I_1$. Multiply the first expression by ω and the second by k (assuming without loss of generality the ω and k are non-zero) and add to obtain

$$\omega = \langle \omega dA(Z) + kdB(Z), Z_1\rangle = \langle DS(Z), Z_1\rangle = \frac{\partial S}{\partial I_1}$$

where the second equality follows from (1.3). This proves the first expression in (1.8) and the second expression in (1.8) follows by a similar argument. The relations in (1.8) also highlight the fact that S can be considered, when evaluated at a critical point, as a function of I_1 and I_2. Another way of seeing this is to note that a solution of (1.3) is of the form $\widehat{Z}(\omega, k)$. But substitution of $\widehat{Z}$ into the constraints, $A(\widehat{Z}(\omega, k)) = I_1$ and $B(\widehat{Z}(\omega, k)) = I_2$, and inversion, results in $\omega(I_1, I_2)$, $k(I_1, I_2)$ and $S(I_1, I_2)$. This all looks much like the operations on integrable 2-DOF Hamiltonian systems (Lichtenberg & Lieberman [1983, p. 311]). In fact, a relation between the sign of the KAM determinant and the linear stability of plane waves can be formulated as follows. The KAM determinant is the determinant of the Hessian of the Hamiltonian function with respect to the actions,

$$\Delta_K = \begin{vmatrix} \frac{\partial^2 S}{\partial I_1^2} & \frac{\partial^2 S}{\partial I_1 \partial I_2} \\ \frac{\partial^2 S}{\partial I_1 \partial I_2} & \frac{\partial^2 S}{\partial I_2^2} \end{vmatrix} = \begin{vmatrix} \frac{\partial \omega}{\partial I_1} & \frac{\partial \omega}{\partial I_2} \\ \frac{\partial k}{\partial I_1} & \frac{\partial k}{\partial I_2} \end{vmatrix} . \tag{1.12}$$

One of the hypotheses of the KAM theorem is that the above determinant must be non-zero for the persistence of invariant tori (Arnold et al. [1988, p. 183]). The non-zero determinant insures that the two frequencies (or in the present case the wavenumber and frequency) are not functionally related (Lichtenberg & Lieberman [1983, p. 162]). The sign of Δ_K has no significance for the KAM theory; it simply gives curvature information about the energy surfaces. However, for the case of periodic plane waves we obtain the remarkable result that the sign of the KAM determinant is related to linear stability. We prove the following result for the generalized NLS equation in Section 2.

Theorem (Plane wave instability). *Let $S(I_1, I_2)$ be evaluated at a plane wave solution of NLS-V. Suppose S (or $-S$) is a convex function of (I_1, I_2) then the plane wave is linearly unstable. If it is a saddle point of S; that is, $\Delta_K < 0$, then the plane wave is linearly stable.*

The result as stated is counter-intuitive since one would expect the convex case to be associated with stability! The result is stated within the context of NLS-V since a proof can be given easily via an explicit calculation. We expect that the result can be generalized but the proof will be considerably longer. Some results in this direction are given in Section 4.

The discussion so far has been limited to plane waves. But the variational principle and subsequent analysis only assumes periodicity in space and time so includes a much larger class of waves. For example; periodic standing waves: these are waves that are periodic in space and time but don't move – have zero momentum. Standing waves are more difficult to analyze since there is no frame of reference in which they are steady (time-independent). In Section 4 the coupled problem of standing waves and travelling waves of the nonlinear Klein-Gordon equation is treated along the lines discussed so far for travelling waves.

2 Plane waves of NLS-V and the KAM determinant

In this section the theory described in the introduction is applied to plane-wave solutions of the nonlinear Schrödinger equation with arbitrary potential,

$$i\frac{\partial\psi}{\partial t} + \frac{\partial^2\psi}{\partial x^2} + V'(|\psi|^2)\psi = 0\,. \tag{2.1}$$

Plane wave solutions are of the form $\psi(x,t) = Re^{i(kx-\omega t)}$. Such solutions are completely elementary and are well-studied (see Newton & Keller [1987], Infeld & Rowlands [1990] and references therein). Our purpose here is to use NLS-V as a simple example, where everything can be computed explicitly, to illustrate a structure that is reminiscent of invariant toroids of classical finite-dimensional Hamiltonian systems. Our main interest is in generalizing and applying this result to systems which are not integrable in any sense and where the plane wave solutions are non-trivial (such as the water-wave problem).

Substituting the plane wave solution $\psi(x,t) = Re^{i(kx-\omega t)}$ into (2.1) results in the nonlinear dispersion relation

$$\omega = k^2 - V'(u) \quad with \quad u = |R|^2\,. \tag{2.2}$$

The linear stability of this plane wave can be determined explicitly following Newton & Keller [1987]. We give the result here as parts of it will be needed in the later development. Let $\psi(x,t) = (R + B(x,t))e^{i(kx-\omega t)}$; then substitution into (2.1) and linearization about the basic state results in the following linear equation for B,

$$iB_t + 2ikB_x + B_{xx} + uV''(u)B + V''(u)R^2\overline{B} = 0\,. \tag{2.3}$$

The general solution of (2.3) is easily seen to be

$$B(x,t) = B_1e^{i(\alpha x-\Omega t)} + B_2e^{-i(\alpha x-\Omega^* t)} \qquad with \quad (B_1, B_2) \in \mathbb{C}^2 \tag{2.4}$$

where Ω^* is the complex conjugate of Ω and $\alpha \in \mathbb{R}$. Substituting (2.4) into (2.3) results in the linear system

$$\begin{bmatrix} \Omega - \alpha^2 - 2k\alpha + uV''(u) & R^2V''(u) \\ \overline{R}^2V''(u) & -\Omega - \alpha^2 + 2k\alpha + uV''(u) \end{bmatrix} \begin{pmatrix} B_1 \\ \overline{B}_2 \end{pmatrix} = 0\,.$$

Setting to zero the determinant results in the characteristic equation for the stability exponent Ω,

$$\Omega^2 - 4k\alpha\Omega + 2V''(u)[2k^2/V''(u) + u]\alpha^2 - \alpha^4 = 0 \tag{2.5}$$

so

$$\Omega = 2\alpha k \pm \alpha\sqrt{\alpha^2 - 2uV''(u)}\,. \tag{2.6}$$

A plane wave is linearly stable if, for all $\alpha \in \mathbb{R}$, the exponent $\Omega \in \mathbb{R}$ and unstable if for some $\alpha \in \mathbb{R}$, $Im(\Omega) \neq 0$. It is clear from (2.6) that the sign of $V''(u)$ determines stability: if $V''(u) < 0$ the plane wave is linearly stable whereas if $V''(u) > 0$ there exists a band of wavenumbers in α-space such that exponentially growing solutions of (2.4) exist – hence linear instability. Newton & Keller [1987, p. 961] give a number of examples with different potentials. The most well-known example is the cubic nonlinear Schrödinger equation, obtained by taking $V(u) = \frac{1}{2}\epsilon u^2$, for which the plane wave is stable for $\epsilon = -1$ (defocussing nonlinearity) and unstable for $\epsilon = +1$ (focussing nonlinearity). This is the classical approach to the study of plane waves of NLS-V. We now consider the problem from the point of view of a space-time Hamiltonian structure.

The evolution equation (2.1) can be written as an infinite-dimensional Hamiltonian system in the following way

$$\psi_t = \mathcal{J} \cdot \frac{\delta H}{\delta \overline{\psi}} \qquad \textit{with} \quad \mathcal{J} = -i \tag{2.7}$$

and

$$H(\psi, \overline{\psi}) = \int_{x_1}^{x_2} \left[\, |\psi_x|^2 - V(|\psi|^2) \,\right] dx \,.$$

However, this Hamiltonian structure is not suitable for the space-time structure that leads to the plane wave variational principle. Essentially, the space derivative terms need to be subtracted from $H(\overline{\psi}, \psi)$ to form a spatial action.

Therefore new coordinates are introduced, defined by

$$Z = \begin{pmatrix} q_1 \\ q_2 \\ p_1 \\ p_2 \end{pmatrix} \stackrel{\text{def}}{=} \begin{pmatrix} \mathrm{Re}(\psi) \\ -\mathrm{Im}(\psi) \\ -\mathrm{Re}(\psi_x) \\ \mathrm{Im}(\psi_x) \end{pmatrix},$$

in terms of which equation (2.1) can be rewritten in the form

$$\mathbf{M} Z_t + \mathbf{K} Z_x = dS(Z) \tag{2.8}$$

with

$$\mathbf{M} = \begin{pmatrix} \mathbf{J} & \mathbf{0} \\ \mathbf{0} & \mathbf{0} \end{pmatrix}, \qquad \mathbf{K} = \begin{pmatrix} \mathbf{0} & -\mathbf{I} \\ \mathbf{I} & \mathbf{0} \end{pmatrix} \quad \textit{with} \quad \mathbf{J} = \begin{pmatrix} 0 & 1 \\ -1 & 0 \end{pmatrix}.$$

The function $S(Z)$ is the S-Hamiltonian function and is given by

$$S(Z) = -\tfrac{1}{2}(p_1^2 + p_2^2) - \tfrac{1}{2} V(q_1^2 + q_2^2)\,. \tag{2.9}$$

The term S-Hamiltonian function is used to distinguish it from the Hamiltonian in a standard system such as (2.7). The symbol S doesn't have any particular significance other than the fact that the S-Hamiltonian is related to the momentum flux which is often given the symbol S. The role of conservation laws in the construction of (2.8) is discussed in Section 3.

The operators $\mathbf{M}Z_t$ and $\mathbf{K}Z_x$ are the first variations of the time and space actions respectively. This can be seen as follows; let

$$\widehat{A} = q_1 \frac{\partial q_2}{\partial t} \quad \textit{and} \quad \widehat{B} = p_1 \frac{\partial q_1}{\partial x} + p_2 \frac{\partial q_2}{\partial x} \tag{2.10}$$

be the time and space action densities. The respective actions are obtained by integrating over x and t. Now

$$\begin{aligned} d\widehat{A} &= \frac{\partial q_2}{\partial t} dq_1 - \frac{\partial q_1}{\partial t} dq_2 + \frac{\partial}{\partial t} \langle q_1 dq_2 \rangle \\ d\widehat{B} &= \langle q_x, dp \rangle - \langle p_x, dq \rangle + \frac{\partial}{\partial x} \langle p, dq \rangle \end{aligned}$$

where $\langle \cdot, \cdot \rangle$ is a real inner product on $\mathbb{R}^2$. Taking suitable fixed endpoint conditions on q we find

$$dA(Z) = \mathbf{M} Z_t \quad \textit{and} \quad dB(Z) = \mathbf{K} Z_x \,.$$

Therefore the appropriate functionals for the plane wave variational principle are S in (2.9) and A, B with densities given in (2.10). Introduce the averaging operator

$$\oint_{\mathbb{T}^2} S\,dxdt = \left(\frac{1}{2\pi}\right)^2 \int_0^{2\pi}\int_0^{2\pi} S\,dxdt\,.$$

Then writing $A = \oint_{\mathbb{T}^2} \widehat{A}\,dxdt$ and $B = \oint_{\mathbb{T}^2} \widehat{B}\,dxdt$ plane-wave solutions of NLS-V are critical points of S on level sets of A and B, or equivalently, with ω, k as Lagrange multipliers, critical points of the unconstrained functional

$$\mathcal{F} = S - \omega A - kB\,. \tag{2.11}$$

In general the set of all possible critical points of the functional (2.11) consists of functions of the form $\sum a_{mn} e^{i(mx+nt)}$ with sum over $(m,n) \in \mathbb{Z}^2$. We restrict attention to the simplest class of plane waves which take the form

$$\begin{aligned} q &= R_\theta \hat{q} \\ p &= R_\theta \hat{p} \\ \text{where}\quad R_\theta &= \begin{pmatrix} \cos\theta & \sin\theta \\ -\sin\theta & \cos\theta \end{pmatrix} \quad \text{and} \quad \theta = x - t\,. \end{aligned}$$

With this expression for Z the functions S, A, B averaged over $\mathbb{T}^2$ become

$$\begin{aligned} S &= -\tfrac{1}{2}|\hat{p}|^2 - \tfrac{1}{2}V(|\hat{q}|^2) \\ A &= \tfrac{1}{2}|\hat{q}|^2 \\ B &= \langle \hat{p}, \mathbf{J}\hat{q}\rangle\,. \end{aligned}$$

So

$$\mathcal{F} = -\tfrac{1}{2}|\hat{p}|^2 - \tfrac{1}{2}V(|\hat{q}|^2) - \tfrac{1}{2}\omega|\hat{q}|^2 - k\langle \hat{p}, \mathbf{J}\hat{q}\rangle$$

and

$$\begin{aligned} 0 &= \frac{\delta\mathcal{F}}{\delta\hat{p}} = -\hat{p} - k\mathbf{J}\hat{q} \\ 0 &= \frac{\delta\mathcal{F}}{\delta\hat{q}} = -V'(|\hat{q}|^2)\hat{q} - \omega\hat{q} + k\mathbf{J}\hat{p}\,. \end{aligned}$$

Setting $\delta\mathcal{F} = 0$ requires $\hat{p} = -k\mathbf{J}\hat{q}$ and

$$\left[\omega - k^2 + V'(|\hat{q}|^2)\right]\hat{q} = 0 \tag{2.12}$$

which recovers the nonlinear dispersion relation (2.2).

Setting $\delta\mathcal{F} = 0$ is a necessary condition for a minimum of the constrained critical point problem: minimize S on level sets of A and B. It is natural to inquire about the second derivative test. The second variation of $\mathcal{F}$ is

$$L \stackrel{def}{=} d^2\mathcal{F} = d^2S - \omega d^2A - kd^2B\,.$$

The usually quoted sufficient condition (neglecting technicalities associated with infinite dimensional spaces) for a critical point of $\mathcal{F}$ to be a minimum of $S|_{A,B}$ is that

$$\langle y, Ly\rangle > 0 \quad \text{for all } y \text{ such that } \langle dA, y\rangle = \langle dB, y\rangle = 0\,.$$

The problem here is that L is indefinite (a countable number of positive *and* negative eigenvalues) on the space of functions periodic in x and t. Therefore there is no hope that a plane wave will be a minimum of the S-Hamiltonian. But the plane wave variational principle contains other information.

Maddocks [1992] has proposed an alternative criteria for the second derivative test for constrained critical points using the sensitivity matrix. The sensitivity matrix is the Jacobian of partial derivatives with respect to the Lagrange multipliers. Maddocks' theory does not apply to the plane wave problem – again because L is indefinite. But it turns out that the sensitivity matrix can be related directly to the stability exponents of the plane wave problem! This argument is a generalization of the argument used in Bridges [1991] for stability of periodic solutions of Hamiltonian systems.

The sensitivity matrix for the plane wave problem is defined by

$$\mathbf{M}_S = \begin{pmatrix} \frac{\partial A}{\partial \omega} & \frac{\partial A}{\partial k} \\ \\ \frac{\partial B}{\partial \omega} & \frac{\partial B}{\partial k} \end{pmatrix} .$$

At a critical point (a plane wave) the action functionals A and B reduce to

$$A = \tfrac{1}{2}|\hat{q}|^2 \quad \textit{and} \quad B = -k|\hat{q}|^2$$

with $|\hat{q}|^2$ defined as a function of ω and k using (2.12). In other words

$$\frac{\partial}{\partial \omega}|\hat{q}|^2 = -\frac{1}{V''(u)} \quad \textit{and} \quad \frac{\partial}{\partial k}|\hat{q}|^2 = \frac{2k}{V''(u)} .$$

Therefore the sensitivity matrix evaluated at a plane wave is

$$\mathbf{M}_S = \begin{pmatrix} -\frac{1}{2V''(u)} & \frac{k}{V''(u)} \\ \frac{k}{V''(u)} & -|\hat{q}|^2 - \frac{2k^2}{V''(u)} \end{pmatrix} .$$

A direct connection between the determinant of the sensitivity matrix and the linear stability of a plane wave is made by direct appeal to the characteristic equation (2.5). First note that

$$\Delta_S = det(\mathbf{M}_S) = \frac{|\hat{q}|^2}{2V''(u)} .$$

Now, multiply the characteristic equation (2.5) by -1/2$V''(u)$ to obtain

$$-\frac{1}{2V''(u)}\Omega^2 + \frac{2k\alpha}{V''(u)}\Omega - \left(|\hat{q}|^2 + \frac{2k^2}{V''(u)}\right)\alpha^2 + \frac{\alpha^4}{2V''(u)} = 0 .$$

But appeal to the entries of the sensitivity matrix shows that this is in fact

$$\frac{\partial A}{\partial \omega}\Omega^2 + (\frac{\partial A}{\partial k} + \frac{\partial B}{\partial \omega})\Omega\alpha + \frac{\partial B}{\partial k}\alpha^2 - \frac{\partial A}{\partial \omega}\alpha^4 = 0$$

from which it follows that

$$\Omega = \frac{-A_k\alpha \pm \alpha\sqrt{A_\omega^2\alpha^2 - \Delta_S}}{A_\omega}$$

(using the fact that the sensitivity matrix is symmetric). Therefore, it is precisely the sign of Δ_S that determines linear stability of the plane wave. We have proved the following.

Theorem. *Let $\psi_o(x,t)$ be a periodic plane wave solution of $NLS - V$ with time and space actions (A,B) and frequency-wavenumber (ω,k). The wave is linearly stable if Δ_S is negative and linearly unstable if Δ_S is positive.*

It remains to prove the Theorem stated in the introduction relating instability results to the sign of the KAM determinant – expressed in terms of the S-Hamiltonian function $S(I_1,I_2)$. Since $2I_1 = |\hat{q}|^2$ it follows, using (2.9) and the fact that $\hat{p} = -k\mathbf{J}\hat{q}$, that S evaluated at a plane wave reduces to

$$\begin{aligned} S &= -\tfrac{1}{2}|\hat{p}|^2 - \tfrac{1}{2}V(|\hat{q}|^2) \\ &= -\tfrac{1}{2}k^2|\hat{q}|^2 - \tfrac{1}{2}V(|\hat{q}|^2) \\ &= -k^2 I_1 - \tfrac{1}{2}V(2I_1)\,. \end{aligned}$$

In the above expression k is considered to be a function of I_1 and I_2; the explicit dependence is easily seen to be

$$k(I_1,I_2) = -\frac{I_2}{2I_1}\,.$$

Therefore, a straightforward computation shows that

$$d^2S(I) = \begin{bmatrix} \frac{\partial^2 S}{\partial I_1^2} & \frac{\partial^2 S}{\partial I_1 \partial I_2} \\ \frac{\partial^2 S}{\partial I_1 \partial I_2} & \frac{\partial^2 S}{\partial I_2^2} \end{bmatrix} = \begin{bmatrix} \frac{\partial \omega}{\partial I_1} & \frac{\partial \omega}{\partial I_2} \\ \frac{\partial k}{\partial I_1} & \frac{\partial k}{\partial I_2} \end{bmatrix} = \begin{bmatrix} 2k\frac{\partial k}{\partial I_1} - 2V''(u) & \frac{\partial k}{\partial I_1} \\ \frac{\partial k}{\partial I_1} & \frac{\partial k}{\partial I_2} \end{bmatrix},$$

so,

$$\Delta_K = \det\left[d^2S(I)\right] = -2V''(u)\frac{\partial k}{\partial I_2} = V''(u)/I_1 = \frac{1}{\Delta_S}\,.$$

Since $V''(u) < 0$ implies stability we have proved the Theorem stated in the introduction – almost. The explicit relation between Δ_K and the stability exponents is obtained by noting that d^2S is just the inverse of the sensitivity matrix $\mathbf{M}_S$!

3 Conservation laws and the S-Hamiltonian function

The most straightforward way to determine the S-Hamiltonian function is to use the Legendre transform. This idea can be seen by example using the nonlinear Klein-Gordon equation

$$\frac{\partial^2 \phi}{\partial x^2} - \frac{\partial^2 \phi}{\partial t^2} = V'(\phi)\,. \tag{3.1}$$

Let $v = \phi$ and $u = \phi_t$ then (3.1) can be written as a Hamiltonian evolution equation in the standard way

$$\frac{\partial}{\partial t}\begin{pmatrix} v \\ u \end{pmatrix} = \begin{bmatrix} 0 & 1 \\ -1 & 0 \end{bmatrix}\begin{pmatrix} \delta H/\delta v \\ \delta H/\delta u \end{pmatrix}$$

with

$$H(v,u) = \int_{x_1}^{x_2} \left[\tfrac{1}{2}u^2 + \tfrac{1}{2}v_x^2 + V(v)\right]\,dx\,.$$

To determine the space-time formulation, where no x-derivatives appear in the S-Hamiltonian function, we go back to (3.1), note the Lagrangian function and perform a Legendre transform. The Lagrangian density for the nonlinear Klein-Gordon equation is (Whitham [1974, p. 491])

$$\mathcal{L}(\phi,\phi_x,\phi_t) = \tfrac{1}{2}\phi_x^2 - \tfrac{1}{2}\phi_t^2 + V(\phi)\,.$$

The governing equation (3.1) is recovered in the usual way from

$$\frac{\partial}{\partial t}\left(\frac{\partial \mathcal{L}}{\partial \phi_t}\right) + \frac{\partial}{\partial x}\left(\frac{\partial \mathcal{L}}{\partial \phi_x}\right) - \frac{\partial \mathcal{L}}{\partial \phi} = 0\,.$$

Now let $v = \phi$,

$$\begin{aligned} u &= -\frac{\partial \mathcal{L}}{\partial \phi_t} = \phi_t \\ w &= \frac{\partial \mathcal{L}}{\partial \phi_x} = \phi_x\,. \end{aligned}$$

Then the spatial Hamiltonian is obtained from

$$\begin{aligned} S &= -u\phi_t + w\phi_x - \mathcal{L} \\ &= \tfrac{1}{2}w^2 - \tfrac{1}{2}u^2 - V(v)\,. \end{aligned}$$

It then follows that the space-time form of the evolution equation (3.1) is obtained from the first variation of

$$\mathcal{F} = \int_{t_1}^{t_2}\int_{x_1}^{x_2} [vu_t + wv_x - S(u,v,w)]\,dxdt$$

resulting in

$$\mathbf{M}Z_t + \mathbf{K}Z_x = dS(Z) \tag{3.2}$$

with $Z = (u, v, w)$ and

$$\mathbf{M} = \begin{pmatrix} 0 & -1 & 0 \\ 1 & 0 & 0 \\ 0 & 0 & 0 \end{pmatrix}, \quad \mathbf{K} = \begin{pmatrix} 0 & 0 & 0 \\ 0 & 0 & -1 \\ 0 & 1 & 0 \end{pmatrix} \quad \textit{and} \quad dS = \begin{pmatrix} -u \\ -V'(v) \\ w \end{pmatrix}.$$

A lot of the information about the structure of the form $\mathbf{M}Z_t + \mathbf{K}Z_x = dS(Z)$ is also contained in the conservation laws; particularly the energy and momentum conservation laws. No physical meaning is attached to the terms energy and momentum; these are simply the conservation laws that follow via Noether's theorem from the symmetries: time-translation invariance generates the energy conservation law and the space translation invariance generates the momentum conservation law. There are often additional conservation laws in particular problems and they must also be taken into consideration.

For NLS-V it is easily found that the energy and momentum conservation laws are (following Bridges [1992a, Section 2]),

$$\begin{aligned} \frac{\partial I}{\partial t} + \frac{\partial S_{xx}}{\partial x} &= 0 \qquad \text{(momentum conservation)} \\ \frac{\partial E}{\partial t} + \frac{\partial F}{\partial x} &= 0 \qquad \text{(energy conservation)} \end{aligned}$$

with

$$I = \mathrm{Im}(\psi\overline{\psi}_x)\,, \quad S_{xx} = \mathrm{Im}(\overline{\psi}\psi_t) - |\psi_x|^2 - V(|\psi|^2)$$

and

$$E = |\psi_x|^2 - V(|\psi|^2)\,, \quad F = -2\mathrm{Re}(\psi_t\overline{\psi}_x)\,.$$

The momentum flux contains the information about the S-Hamiltonian function. In fact,

$$\begin{aligned} S_{xx} &= -A + S \\ E &= -B + S \end{aligned}$$

with $A = -\mathrm{Im}(\overline{\psi}\psi_t)$, so the space action can be expressed as

$$B = A + S_{xx} - E\,.$$

In words, the densities of (space)-action minus (time)-action equals the densities of momentum flux minus energy. The functional A is the (time)-action for the basic Hamiltonian evolution equation (2.7) since

$$\frac{\delta A}{\delta \overline{\psi}} = i\psi_t$$

so (2.7) is recovered from

$$\frac{\delta A}{\delta \overline{\psi}} = \frac{\delta H}{\delta \overline{\psi}}$$

which is the first variation of the usual least action functional with density $H - A$.

The relationship of the momentum flux to S-Hamiltonian function is closely related to the energy-momentum flux formulation for relative equilibria (Bridges [1992a,b]). For example the set of equations (2.8) under the transformation $x \mapsto x - ct$ becomes

$$[\mathbf{K} - c\mathbf{M}]Z_x = dS(Z) \tag{3.3}$$

which is a two-DOF spatial Hamiltonian system with spatial Hamiltonian $S(Z)$. The skew-symmetric operator $\mathbf{K} - c\mathbf{M}$ generates the symplectic form for the spatial Hamiltonian system (3.4):

$$\omega = dq_1 \wedge dp_1 + dq_2 \wedge dp_2 - c\,dq_1 \wedge dq_2\,.$$

A linear change of coordinates (which also transforms $S(Z)$ to the momentum flux relative to a moving frame) will take the symplectic form to the standard one (Bridges [1992a]).

4 Space and time periodic waves – standing waves

Although the emphasis so far has been on plane waves the theory is more general, including all space and time periodic states. The most interesting class of such states, other than plane waves, is standing waves. Standing waves are space and time periodic but have zero momentum. At the linear level they consist of the superposition of two plane waves of the same amplitude travelling in opposite directions. In this section the theory is sketched for the case where both standing waves and travelling waves are included in the analysis. The nonlinear Klein-Gordon equation is used as an example; mainly because there is a great deal of information available in the literature about the stability of plane waves of the Klein-Gordon equation (although very little about stability of standing waves!).

Standing waves are more difficult to analyze than travelling waves because they are truly time-dependent (there does not exist a frame of reference such that they are steady). They fit nicely into the variational structure discussed in the introduction but the stability theory for standing waves is incomplete. There is very little in the literature on the stability of standing waves of Klein-Gordon and related equations. Gibbon & Knobloch [1991] have proposed that the stability question for standing waves (called counter-propagating waves there) can be addressed by deriving non-local coupled NLS equations but do not present any stability results. The present results on stability of standing waves are speculative – suggesting a role for the sign of the KAM determinant – and a definitive statement will have to await a careful proof relating the information here (sign of the KAM determinant or sensitivity matrix determinant) to stability exponents for time-dependent standing waves.

On the other hand there is an abundance of theories for the stability of plane wave solutions of the nonlinear Klein-Gordon equation. By comparison with known results it will be shown that a positive (negative) sign of the KAM determinant predicts the linear instability (stability) of plane waves of Klein-Gordon.

Another aspect of this section is that the theory is presented in a context where the exact solution is not known and a formal double Fourier series expansion, which should be useful in other contexts, is used to obtain an approximate, small norm, expression for waves. The convergence question for the Fourier series is not considered. It is very likely that small divisor difficulties will enter when both travelling waves and standing waves are included in the analysis. When only travelling waves are considered the small divisor difficulty can generally be eliminated by going to a moving frame (this is a bit like the case of invariant toroids in finite-dimensional Hamiltonian systems where restriction to periodic states results in convergent series without small divisors).

The space-time formulation for the Klein-Gordon equation is given in (3.2)-(3.3). Linearization about the trivial state results in

$$\begin{bmatrix} 0 & -1 & 0 \\ 1 & 0 & 0 \\ 0 & 0 & 0 \end{bmatrix} \begin{pmatrix} u \\ v \\ w \end{pmatrix}_t + \begin{bmatrix} 0 & 0 & 0 \\ 0 & 0 & -1 \\ 0 & 1 & 0 \end{bmatrix} \begin{pmatrix} u \\ v \\ w \end{pmatrix}_x = \begin{bmatrix} -1 & 0 & 0 \\ 0 & -1 & 0 \\ 0 & 0 & 1 \end{bmatrix} \begin{pmatrix} u \\ v \\ w \end{pmatrix} . \tag{4.1}$$

Corresponding to the dispersion relation $\omega^2 = k^2+1$ equation (4.1) has the following basic solution

$$\begin{pmatrix} u \\ v \\ w \end{pmatrix} = 2\,\mathrm{Re} \left\{ \begin{pmatrix} -i\omega \\ 1 \\ ik \end{pmatrix} a\, e^{i(kx-\omega t)} + \begin{pmatrix} i\omega \\ 1 \\ ik \end{pmatrix} b\, e^{i(kx+\omega t)} \right\} \tag{4.2}$$

where (a, b) are arbitrary complex numbers. If $b = 0$ but $a \neq 0$ (or $a = 0$ and $b \neq 0$) the basic state is a plane wave and if $|a| = |b|$ the basic state is a standing wave.

Writing $Z = (u, v, w)$ the general solution of the periodic nonlinear problem has the form

$$Z(x, t) = \sum_{(m,n)\in\mathbb{Z}^2} Z_{mn} e^{i(mx+nt)} .$$

Our goal here is to take a finite Fourier expansion sufficient to obtain a local theory valid for all small space and time periodic states and apply the $S-A-B$ variational principle for space-time periodic waves. With a slightly different basis for the Fourier series (which separates right and left running waves) we take the following finite Fourier expansion for Z,

$$
\begin{aligned}
Z(x,t) \stackrel{\text{def}}{=} \begin{pmatrix} u(x,t) \\ v(x,t) \\ w(x,t) \end{pmatrix} &= A_o \begin{pmatrix} 0 \\ 1 \\ 0 \end{pmatrix} + 2\mathrm{Re}\left\{ a \begin{pmatrix} i\omega \\ 1 \\ ik \end{pmatrix} e^{i(x-t)} + b \begin{pmatrix} -i\omega \\ 1 \\ ik \end{pmatrix} e^{i(x+t)} \right. \\
&\quad + A_2 \begin{pmatrix} 2i\omega \\ 1 \\ 2ik \end{pmatrix} e^{2i(x-t)} + B_2 \begin{pmatrix} -2i\omega \\ 1 \\ 2ik \end{pmatrix} e^{2i(x+t)} \\
&\quad \left. + A_3 \begin{pmatrix} 0 \\ 1 \\ 2ik \end{pmatrix} e^{2ix} + B_3 \begin{pmatrix} 2i\omega \\ 1 \\ 0 \end{pmatrix} e^{2it} \right\} + \cdots .
\end{aligned} \tag{4.3}
$$

To apply the $S - A - B$ variational principle, introduce integrals for $S(Z)$, $A(Z)$ and $B(Z)$ averaged over $\mathbb{T}^2$,

$$
\begin{aligned}
S &= \oint_{\mathbb{T}^2} \left\{ \tfrac{1}{2}w^2 - \tfrac{1}{2}u^2 - V(v) \right\} dxdt \\
A &= \oint_{\mathbb{T}^2} \{ v u_t \} \, dxdt \\
B &= \oint_{\mathbb{T}^2} \{ w v_x \} \, dxdt
\end{aligned} \tag{4.4}
$$

where $V(v)$ is some smooth function with Taylor expansion

$$
V(v) = \tfrac{1}{2}v^2 + \tfrac{1}{3}\alpha v^3 + \tfrac{1}{4}\beta v^4 + \cdots .
$$

Substitution of (4.3) into (4.4) results in

$$
\begin{aligned}
A &= -2\omega(|a|^2 + |b|^2) - 8\omega(|A_2|^2 + |B_2|^2 + |B_3|^2) + \cdots \\
B &= 2k(|a|^2 + |b|^2) + 8k(|A_2|^2 + |B_2|^2 + |A_3|^2) + \cdots \\
S &= (k^2 - \omega^2 - 1)(|a|^2 + |b|^2) + (4k^2 - 4\omega^2 - 1)(|A_2|^2 + |B_2|^2) - \tfrac{3}{2}\beta(|a|^4 + |b|^4) \\
&+ (4k^2 - 1)|A_3^2| - (4\omega^2 + 1)|B_3^2| - \tfrac{1}{2}A_o^2 - 2\alpha A_o(|a|^2 + |b|^2) - 6\beta |a|^2 |b|^2 \\
&- \alpha(a^2 \overline{A}_2 + \overline{a}^2 A_2 + b^2 \overline{B}_2 + \overline{b}^2 B_2 + 2ab\overline{A}_3 + 2\overline{a}\overline{b}A_3 + 2a\overline{b}B_3 + 2\overline{a}b\overline{B}_3) + \cdots .
\end{aligned}
$$

The small-norm space-time periodic waves of the nonlinear Klein-Gordon equation are obtained as critical points of the unconstrained functional

$$
\mathcal{F} = S - \omega A - kB .
$$

Setting the derivative of $\mathcal{F}$ with respect to A_3, B_3, A_2, B_2 and A_o to zero results in

$$
\begin{aligned}
A_3 &= -\frac{2\alpha}{1 + 4k^2} ab + \cdots \qquad & B_3 &= \frac{2\alpha}{4\omega^2 - 1} \overline{a} b + \cdots \\
A_2 &= \frac{\alpha}{4\omega^2 - 4k^2 - 1} a^2 + \cdots \qquad & B_2 &= \frac{\alpha}{4\omega^2 - 4k^2 - 1} b^2 + \cdots \\
A_o &= -2\alpha(|a|^2 + |b|^2) + \cdots .
\end{aligned}
$$

Back substituting into $\mathcal{F}$ results in the reduced functional

$$\begin{aligned}\mathcal{F} &\stackrel{def}{=} S - \omega A - kB \qquad (4.5)\\ &= (\omega^2 - k^2 - 1)(|a|^2 + |b|^2) + \tfrac{1}{2} f_1 |a|^4 + f_2 |a|^2 |b|^2 + \tfrac{1}{2} f_1 |b|^4 + \cdots\end{aligned}$$

where

$$f_1 = 4\alpha^2 - 3\beta - \frac{2\alpha^2}{(4\omega^2 - 4k^2 - 1)}$$

$$f_2 = 4\alpha^2 \left(1 - \frac{1}{4\omega^2 - 1} - \frac{1}{1 + 4k^2}\right) - 6\beta$$

from which we obtain the nonlinear dispersion relation

$$\begin{aligned}\delta\mathcal{F}/\delta\bar{a} &= \left[(\omega^2 - k^2 - 1) + f_1|a|^2 + f_2|b|^2 + \cdots\right] a = 0\\ \delta\mathcal{F}/\delta\bar{b} &= \left[(\omega^2 - k^2 - 1) + f_2|a|^2 + f_1|b|^2 + \cdots\right] b = 0\,.\end{aligned} \qquad (4.6)$$

There are two classes of solutions contained in (4.6):

$$b = 0 \quad a \neq 0 \quad \textit{and} \quad |a|^2 = -\frac{(\omega^2 - k^2 - 1)}{f_1} \quad \text{(travelling waves)} \qquad (4.7)$$

and

$$|a| = |b| \quad \textit{and} \quad |a|^2 = -\frac{(\omega^2 - k^2 - 1)}{f_1 + f_2} \quad \text{(standing waves)}\,. \qquad (4.8)$$

We first show that the sign of the KAM determinant determines the linear stability of the TWs and then consider separately the case for standing waves.

For travelling wave solutions of the nonlinear Klein-Gordon equation it is known that a TW with dispersion relation

$$\omega = \omega_o(k) + \Omega_2|a|^2 + \cdots \qquad (4.9)$$

is linearly stable if $\omega_o''(k)\Omega_2 > 0$ and linearly unstable if $\omega_o''(k)\Omega_2 < 0$ (cf. Whitham [1974, p. 490]). From (4.6) the nonlinear dispersion relation for travelling waves is

$$\omega^2 = k^2 + 1 - f_1|a|^2 + \cdots\,.$$

So, with

$$\omega_o = \sqrt{k^2 + 1} \qquad \textit{and} \qquad \Omega_2 = -\frac{f_1}{2\omega_o} \qquad (4.10)$$

the form (4.9) is recovered. We now compare this result with the sign of Δ_K.

The KAM determinant is given by

$$\Delta_K = \begin{vmatrix} \frac{\partial\omega}{\partial I_1} & \frac{\partial\omega}{\partial I_2} \\ \frac{\partial k}{\partial I_1} & \frac{\partial k}{\partial I_2} \end{vmatrix} = \frac{1}{\Delta_S^2} \begin{vmatrix} \frac{\partial I_2}{\partial k} & -\frac{\partial I_1}{\partial k} \\ -\frac{\partial I_2}{\partial \omega} & \frac{\partial I_1}{\partial \omega} \end{vmatrix} = \frac{1}{\Delta_S}$$

where Δ_S is the determinant of the sensitivity matrix. Using the expressions for the space and time actions and (4.7),

$$I_1 = -2\omega|a|^2 + \cdots \quad \textit{and} \quad I_2 = 2k|a|^2 + \cdots\,.$$

Therefore, writing $u = |a|^2$,

$$\begin{aligned}\Delta_S &\stackrel{def}{=} \begin{vmatrix} \frac{\partial I_1}{\partial \omega} & \frac{\partial I_1}{\partial k} \\ \frac{\partial I_2}{\partial \omega} & \frac{\partial I_2}{\partial k} \end{vmatrix} = \begin{vmatrix} -2|a|^2 - 2\omega u_\omega & -2\omega u_k \\ 2ku_\omega & 2|a|^2 + 2ku_k \end{vmatrix} + \cdots \\ &= -4|a|^2 \left\{ k\frac{\partial}{\partial k}|a|^2 + \omega\frac{\partial}{\partial \omega}|a|^2 \right\} + \cdots \\ &= 8\frac{|a|^2}{f_1}(\omega^2 - k^2) + \cdots \\ &= -4\frac{|a|^2}{\Omega_2}\omega_o^2\omega_o'' + \cdots \end{aligned}$$

using (4.10) and the fact that $\omega_o^3\omega_o'' = \omega_o^2 - k^2$. From (4.9) and (4.10) it follows that the travelling plane wave is linearly stable for $\Delta_S < 0$ and unstable for $\Delta_S > 0$.

Performing the same analysis for the standing wave we find that

$$\Delta_S = 32|a|^2\frac{\omega_o^3}{f_1 + f_2}\omega_o'' + \cdots .$$

The only difference between the SW and TW case is the coefficient in the denominator of the leading term of Δ_S. The precise relation between this expression and the linear stability exponents for standing waves will have to await a more complete analysis.

References

Arnold, V.I., Kozlov, V.V. and Neishtadt, A.I. [1988], *Mathematical Aspects of Classical and Celestial Mechanics*, Enc. of Math. Sci., Dynamical Systems III, (ed. V. Arnold), Springer-Verlag.

Bridges, T.J., *Stability of periodic solutions near a collision of eigenvalues of opposite signature*, Math. Proc. Camb. Phil. Soc., **109**, 375-403.

Bridges, T.J. [1992a], *Hamiltonian bifurcations of the spatial structure for coupled nonlinear Schrödinger equations*, Physica D, **57**, 375-394.

Bridges, T.J. [1992b], *Spatial Hamiltonian Structure, energy flux and the water-wave problem*, Proc. Roy. Soc. Lond., **A 439**, 297-315.

Infeld, E. and Rowlands, G. [1990], *Nonlinear waves, Solitons and Chaos*, Cambridge University Press.

Knobloch, E. and Gibbon, J.D. [1991], *Coupled NLS for counter-propagating waves in systems with reflection symmetry*, Phys. Lett., **A154**, 353-56.

Lichtenberg, L. and Lieberman, M.A. [1983] *Regular and Stochastic Motion*, Applied Math. Sciences, **38**, Springer-Verlag.

Maddocks, J. [1992], *On second-order conditions in constrained variational principles*, preprint, University of Maryland.

Newton, P.K. and Keller, J.B. [1987], *Stability of periodic plane waves*, SIAM J. Appl. Math., **47**, 959-64.

Percival, I.C. [1974] *Variational principles for the invariant toroids of classical dynamics*, J. Phys A: Math., Nucl. Gen., **7**, 794-802.

Whitham, G.B. [1974] *Linear and Nonlinear Waves*, John Wiley & Sons, New York.

Fields Institute Communications
Volume 8, 1996

Anharmonic Oscillators in Classical and Quantum Mechanics with Applications to the Perturbed Kepler Problem

Martin Kummer
Mathematics Department
University of Toledo
Toledo, Ohio 43606

1 Introduction

The purpose of this paper is to present a unified approach to the theory of non linearly perturbed harmonic oscillators in classical and quantum mechanics by concentrating on those aspects of the theory that run parallel in both types of mechanics. The physical systems under consideration are modeled by a Hamiltonian which breaks naturally into a unperturbed Hamiltonian corresponding to a finite number of harmonic oscillators and a perturbation which for simplicity we assume to be a polynomial in the position and momentum variables of the oscillators. These systems are not so specialized as one first might think. Indeed, it is well known that the important Kepler problem together with some of its perturbations can be viewed as special cases of such systems, and I do not think that I have to convince anyone about the importance of these problems for astronomy as well as for atomic physics. For many years workers in quantum Hamiltonian mechanics and those in classical mechanics have gone separate ways. There are in the literature many duplications, where a system is treated with one set of methods in the classical realm and with a totally different set of methods in the quantum realm. Indeed, we shall see that one can accomplish a unified approach in both domains, and thereby achieve an economy of thought which is truly amazing. I believe that at a time in which specialization is in the danger to destroy the unity of science it is of utmost importance to economize our thinking by building bridges between different realms of knowledge. Certainly, I am not the first to recognize that on a conceptual level there are common core notions in both the classical and the quantum domain, namely the notion of a group and of normal form of a Hamiltonian.
However, I do not think that it has been generally recognized that the notion of the Poisson algebra is equally important as a link between classical and quantum mechanics. Recall that an abstract Poisson algebra is an associative algebra over C which is also a Lie algebra and the two products $\cdot$ and $[\ ,\ ]$ are intertwined by the relation $[A\cdot B, C] = A\cdot[B, C] + [A, C]\cdot B$. This means that the bracket is a derivation

1991 *Mathematics Subject Classification.* Primary: 34C20; Secondary: 81Q15.

of the associative product. We shall see that in concrete systems these common core notions make even the computational aspects of the problem appear almost identical: Virtually the same normal form which in classical mechanics can be used to search for families of (quasi-) periodic solutions leads to the correct formula for the energy - eigenvalue splitting in the corresponding quantum problem.

When I am talking about normal form in quantum mechanics I do not mean a quantization procedure of the classical socalled Gustavson - Birkhoff normal form. Such an approach is plagued by well known ambiguities.[1] What I mean by quantum normalization is a well defined procedure mimicking the classical procedure in the framework of a non-commutative Poisson algebra instead of a commutative one. A rudimentary form of this quantum normal form I first have presented in the paper Kummer [1971]. It seems to agree with the quantum normal form proposed by Ali [1985]. There are no ambiguities in this procedure that arise from different quantization procedures: In fact in the framework we are working in, there is just one quantization, namely the one which is equivalent to the Stone - von Neumann representation of the commutation relations, although we shall find that the equivalent representation by Bargmann is especially well adapted for our purposes. In all applications for which we have tested our procedure it gives the same results as all the other ad hoc procedures invented to deal with a special set of problems, including the Raleigh - Schrödinger series. The most extensive test I have subjected this procedure to - date is the approximate spectrum of the quadratic Zeeman effect to fourth order in the normalized variables. It contains seventeen coefficients which all agree with the ones obtained by Delande and Gay [1984]. These authors apply a special projection method which works for certain perturbation problems of the Kepler problem within the framework of quantum mechanics.

In order to explain what sort of ideas are involved we first consider classical Hamiltonians of one degree of freedom. A Hamiltonian vector field associated with a function $H(x,p)$ defined on the phase plane (p, x-plane) has the form $X_H = \frac{\partial H}{\partial p}\frac{\partial}{\partial x} - \frac{\partial H}{\partial x}\frac{\partial}{\partial p}$ so that $i_{X_H} dp \wedge dx = -dH$. More informally this means that given the Hamiltonian H and the one form $\theta = pdx$, the associated ODEs are:

$$\dot{x} = \frac{\partial H}{\partial p}, \quad \dot{p} = -\frac{\partial H}{\partial x}. \tag{1}$$

These equations under a transformation $p, x \to P, X$ retain their Hamiltonian form provided that θ in the new variables differs from PdX by an exact differential. For this reason in what follows I will identify (notation $\cong$) one forms differing by an exact differential. Consider first the Hamiltonian of the harmonic oscillator:

$$H_0 = \frac{1}{2}(p^2 + \omega^2 x^2) = \omega N, \quad N = z\bar{z}. \tag{2}$$

Here, we have set

$$z = \sqrt{\frac{\omega}{2}}x - i\frac{1}{\sqrt{2\omega}}p \tag{3}$$

with associated transformation of the one form

[1] See, e.g., Wood and Ali [1985]

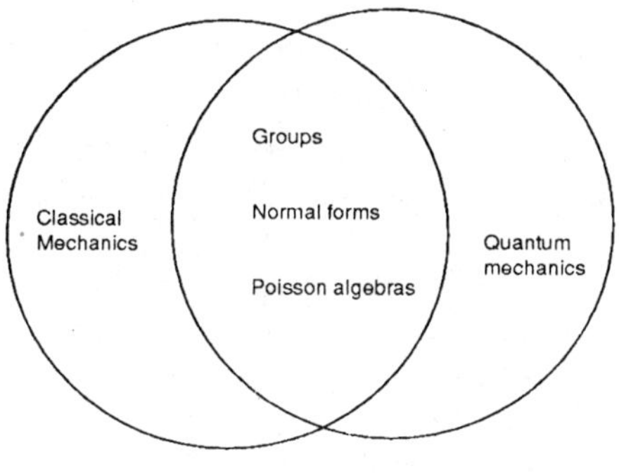

Figure 1 The common core notions of classical and quantum mechanics.

$$\theta = pdx \cong \frac{1}{i}\bar{z}dz \cong Nd\phi. \tag{4}$$

Here $z = \sqrt{N}e^{i\phi}$. The equations of motion for a general one- degree- of -freedom Hamiltonian are

$$\dot{z} = iH_{\bar{z}}, \quad \dot{\bar{z}} = -iH_z \tag{5}$$

or

$$\dot{N} = -H_\phi \quad , \dot{\phi} = H_N. \tag{6}$$

In view of the equation of motion the rate of change of a function defined on the phase plane under the flow induced by H is

$$\dot{f} = f_z\dot{z} + f_{\bar{z}}\dot{\bar{z}} = i[H, f], \tag{7}$$

where [,] is the Poisson bracket

$$[A, B] = \frac{\partial A}{\partial \bar{z}}\frac{\partial B}{\partial z} - \frac{\partial B}{\partial \bar{z}}\frac{\partial A}{\partial z}. \tag{8}$$

In the special case of the harmonic oscillator the equations of motion are:

$$\dot{z} = i\omega z, \quad \dot{\bar{z}} = -i\omega\bar{z} \tag{9}$$

or $\dot{N} = 0$, $\dot{\phi} = \omega$ with solutions $z = z^{(0)}exp(i\omega t)$ and $N = const.$, $\phi = \phi^{(0)} + \omega t$, respectively.

How about quantum mechanics? In order to answer this question let us look at what we have done from an algebraic point of view. Consider now the Poisson algebra generated by three symbols $z, \bar{z}, 1$ subject to the relations

$$[\bar{z}, z] = 1 \tag{10}$$

and all other brackets zero. It is clear that all elements of the algebra generated in this way are polynomials in the two variables $z, \bar{z}$. It is remarkable that the assumption of commutativity of the algebra suffices for the Poisson bracket to have the form of the Poisson bracket as given in Equation (8) (8) (see Theorem 1.2).

On the other hand we obtain quantum theory by simply dropping the assumption of commutativity and interpreting the bracket as commutator. In the language of quantum mechanics this means that z becomes a creation operator and $\bar{z}$ an annihilation operator. Indeed, in terms of the original phase variables our bracket relation corresponds to the Heisenberg commutation relations

$$[p, x] = \frac{1}{i} \tag{11}$$

and the Hamiltonian of the harmonic oscillator is

$$H_0 = \frac{1}{2}(p^2 + \omega^2 x^2) = \omega(N + \frac{1}{2}). \tag{12}$$

In order to obtain the spectrum we employ Bargmann's representation of the commutation relations on his Hilbert space $\mathcal{F}^1$ of analytic functions in the variable z Bargmann [1961]. In this interpretation z is the multiplication operator that replaces $\psi \in \mathcal{F}^1$ by $z\psi$ and $\bar{z}$ symbolizes the partial derivative with respect to z. The inner product in $\mathcal{F}^1$ derives from the norm

$$||\psi||^2 = \frac{1}{2\pi} \int \int exp(-|z|^2)|\psi(z)|^2 dxdp \tag{13}$$

and $\frac{z^n}{\sqrt{n!}}$ is an eigenfunction to the eigenvalue n of the operator N $(n = 0, 1, 2, \dots)$ and since these functions constitute a total orthonormal system in $\mathcal{F}^1$ the spectrum of H_0 is $\omega(n + \frac{1}{2})$ $(n = 0, 1, 2, \dots)$.
In order to study the perturbed problem

$$H_0 + \epsilon V(z, \bar{z}, \epsilon), \tag{14}$$

where the perturbation is a polynomial in z, $\bar{z}$, we resort to the method of normal form which makes sense in both types of mechanics. It is clear that the formal power series in z, $\bar{z}$ form a Lie algebra $\mathcal{L}$ under the bracket. In fact, $\mathcal{L}$ carries a natural grading by the non - negative integers, $\mathcal{L}^{(k)}$ comprising the subspace the polynomials homogeneous of degree k+2. The context of graded Lie algebras is precisely the framework most suitable for a discussion of normalizations of Hamiltonians and vectorfields. This point of view has been emphasized by Baider and Churchill [1988], and Baider [1989].[2] In what follows, we quickly review the procedure of normalization in this framework. Accordingly, we think of $\mathcal{L}$ as an abstract Lie algebra

[2]For other approaches to normal form theory see e.g. Deprit [1969], Moser [1968], van der Meer [1985].

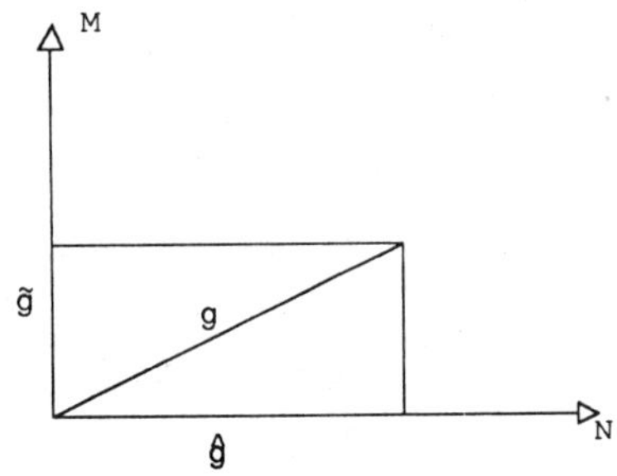

Figure 2 The decomposition of the Lie algebra $\mathcal{L}$.

graded by the nonegative integers. This means that $\mathcal{L}$ admits a direct product decomposition

$$\mathcal{L} = \mathrm{X}_{i=0}^{\infty}\mathcal{L}^{(i)}, \quad \dim\mathcal{L}^{(i)} < \infty \tag{15}$$

where the factors satisfy the bracket relations

$$[\mathcal{L}^{(i)}, \mathcal{L}^{(k)}] \subset \mathcal{L}^{(i+k)}. \tag{16}$$

Instead of writing an element $g \in \mathcal{L}$ as an infinite sequence we write it as a formal sum with coefficients $g^{(k)} \in \mathcal{L}^{(k)}$, i.e., $g = \sum_{k=0}^{\infty} g^{(k)}$. We also use the notation : $\mathcal{L}^{\{k\}} = \mathrm{X}_{i=k}^{\infty}\mathcal{L}^{(i)}$. The goal is to construct order by order an inner automorphism of $\mathcal{L}$ which brings the perturbed Hamiltonian $H^{(0)} + V, \;\; H^{(0)} \in \mathcal{L}^{(0)}, \;\; V \in \mathcal{L}^{\{1\}}$ into a form H that "commutes" with $H^{(0)}$. Stated differently, this means $H \in \mathcal{N} := \ker$ ad$H^{(0)}$. A sufficient condition for this to be possible is the splitting assumption: If $\mathcal{M}$:=Im ad$H^{(0)}$ then $\mathcal{N} \oplus \mathcal{M} = \mathcal{L}$. For any given element $g \in \mathcal{L}$ let $\hat{g}$ be the component of g along $\mathcal{N}$ and $\tilde{g}$ the component along $\mathcal{M}$. (See Figure 2.)

Theorem 1.1 *There exists a formal canonical transformation which converts the given perturbation into normal form with respect to* $H^{(0)}$ *up to any desired order.*

Sketch of a Proof: Note that the expression $exp(adS)C$ for $C \in \mathcal{L}, S \in \mathcal{L}^{\{1\}}$, and with the exponential being thought of as represented by its usual power series, is a well defined element of $\mathcal{L}$. Given $V \in \mathcal{L}^{\{1\}}$ we seek $S_1 \in \mathcal{L}^{(1)}$ so that

$$exp(adS_1)(H^{(0)} + V) = H^{(0)} + H^{(1)} + V_2, \tag{17}$$

where $H^{(1)} \in \mathcal{N}^{(1)}, \;\; V_2 \in \mathcal{L}^{\{2\}}$. By comparing terms of order m on both sides of the last equation we obtain the identity

$$adH^{(0)}S_1 + H^{(1)} = V^{(1)} \tag{18}$$

which is satisfied by setting $H^{(1)}$ equal to $\hat{V}^{(1)}$, i.e., the component of $V^{(1)}$ along $\mathcal{N}$ and letting S_1 be a solution of $adH_0 S_1 = \tilde{V}^{(1)}$ (= component of $V^{(1)}$ along $\mathcal{M}$). We assume inductively that $H^{(k)} \in \mathcal{N}^{(k)}$ as well as $S_k \in \mathcal{L}^{(k)}$ $(k = 1, 2, \ldots, m-1)$ have been constructed in such a way that

$$exp(adS_{m-1}) \ldots exp(adS_1)(H^{(0)} + V) = H^{[m-1]} + V_m \tag{19}$$

with suitable $V_m \in \mathcal{L}^{\{m\}}$. Here and in the following, the abbreviation $H^{[m]} = \sum_{k=0}^{m} H^{(k)}$, etc. is used. Now we seek $S_m \in \mathcal{L}^{(m)}$ in such a way that

$$exp(adS_m)(H^{[m-1]} + V_m) = H^{[m]} + V_{m+1}, \tag{20}$$

where $H^{(m)} \in \mathcal{N}^{(m)}, \quad V_{m+1} \in \mathcal{L}^{\{m+1\}}$. By comparing terms of order m on both sides of the last equation we obtain the identity

$$adH^{(0)} S_m + H^{(m)} = V_m^{(m)} \tag{21}$$

which is satisfied by setting $H^{(m)} = \hat{V}_m^{(m)}$ and letting S_m be a solution of the equation:

$$adH^{(0)} S_m = \tilde{V}_m^{(m)}. \tag{22}$$

q.e.d.

For later use we record the formula

$$H^{(2)} = \frac{1}{2}[S_1, \tilde{V}^{(1)}]\hat{} + \hat{V}^{(2)} \tag{23}$$

which is valid provided we normalize S_1 in such a way that $\hat{S}_1 = 0$. This follows from the fact that in view of the Jacobi identity $\mathcal{N}$ is a subalgebra of $\mathcal{L}$ and $\mathcal{M}$ is invariant under $\mathcal{N}$, i.e.,

$$[\mathcal{N}, \mathcal{N}] \subset \mathcal{N}, \quad [\mathcal{N}, \mathcal{M}] \subset \mathcal{M}. \tag{24}$$

Although the first non-vanishing term in the normal form is uniquely determined, the higher order terms in the normal form are generally not unique: For example, suppose $H^{(1)} \neq 0$ and $H^{(0)} + H^{(1)} + H^{(2)} + \ldots$ is in normal form and $T \in \mathcal{N}^{\{1\}}$. Then $exp(adT)(H^{(0)} + H^{(1)} + H^{(2)} + \ldots) = H^{(0)} + exp(adT)(H^{(1)} + H^{(2)} + \ldots) = H^{(0)} + H^{(1)} + K^{(2)} + \ldots$, where

$$adT^{(1)} H^{(1)} + H^{(2)} = K^{(2)}. \tag{25}$$

Let G^0 be the group of automorphisms of $\mathcal{L}^0$ which fix $H^{(0)}$ and assume that it induces a representation in $\mathcal{N}^{\{1\}}$ so that we can define the semidirect product G of the groups G^0 and $expad\mathcal{N}^{\{1\}}$. Then two normal forms of a given Hamiltonian lie on the same orbit of G. In particular, if $\mathcal{N}$ is commutative and G^0 induces the trivial representation in $\mathcal{N}^{\{1\}}$ then the normal form is uniquely determined.[3] For instance,

[3] For more precise information about the question of uniqueness see Baider and Churchill [1988], Baider [1989].

this is the case for the harmonic oscillator, in which $\mathcal{L}$ = formal power series in z and $\bar{z}$ and $\mathcal{N}$= formal power series in N, $\mathcal{M}$ = those formal power series in z and $\bar{z}$ for which the power of z differs from the one of $\bar{z}$. It turns out that in this case the classical normal form is convergent: Indeed, in this case classical normalization is equivalent to introducing action and angle variables: The normalized Hamiltonian has the form $H(N)$ and the equation of motion

$$\dot{z} = iH(N)_{\bar{z}} = iH'(N)z \tag{26}$$

can be integrated:

$$z = z^{(0)}exp(iH'(N)t). \tag{27}$$

Example:

$$V = ax^3 + bx^4. \tag{28}$$

The dominating term in a normal form corresponding to our Hamiltonian (14) with potential (28) turns out to be

$$[-a^2\frac{15}{4\omega^4} + b\frac{3}{2\omega^2}]N^2, \tag{29}$$

in c.m. and

$$[-a^2\frac{15}{4\omega^4} + b\frac{3}{2\omega^2}]N(N+1) + b\frac{3}{4\omega^2}, \tag{30}$$

in q.m..

Generalization: Birkhoff normal form: $H_0 = \sum_{\nu=0}^{n} \omega_\nu N_\nu$, where the ω_ν are linearly independent over the rationals, $\mathcal{L}$= formal power series in z_ν and $\bar{z}_\nu$ and $\mathcal{N}$ = formal power series in N_ν, $\mathcal{M}$= those formal power series in z_ν and $\bar{z}_\nu$ for which the power of at least one z_ν differs from the one of $\bar{z}_\nu$ ($\nu = 1, 2, \dots, n$). Clearly, if the Birkhoff normal form (in c.m.) is convergent then the system is integrable, the N_ν being a set of n functionally independent commuting integrals. Ito [1989] proved the following converse: Assume the ω_ν are linearly independent over the rationals and the perturbed Hamiltonian $H_0 + V(z, \bar{z})$ admits n functionally independent integrals then the Birkhoff normal form is convergent.

In case there exist linear relations with integral coefficients between the frequencies we speak of resonances. One of the most important resonances which occurs in applications (to celestial mechanics) is the socalled semisimple 1:1 resonance to which we now turn. We consider two harmonic oscillators with the same frequencies and perturbations thereof, i.e., systems described by Hamiltonians of type

$$H = \frac{1}{2}(p_1^2 + q_1^2) + \frac{1}{2}(p_2^2 + q_2^2) + \sum_{k=1} \epsilon^k W_k(p_1, p_2, q_1, q_2). \tag{31}$$

Here W_k are real polynomials in the indicated variables and ϵ is a coupling constant. The most important case in classical mechanics is the case in which W_k is homogeneous of degree k+2 and ϵ is a parameter which is artificially introduced by the method of stretching the variables. However, in view of the fact that we want to study this Hamiltonian in quantum mechanics later, and stretching of variables does not makes sense in quantum mechanics, we may as well consider the more general case from the outset. Of course, the one form corresponding to the Hamiltonian in Eq. (31) is $\sum_{k=1}^{k=2} p_k dq_k \simeq \frac{1}{i}\sum_{k=1}^{k=2} \bar{z}_k dz_k$. Here, the z- variables are defined in analogy to the one degree of freedom case as follows:

$$z_k = \frac{1}{\sqrt{2}}(q_k - ip_k)\ (k = 1, 2). \tag{32}$$

The Hamiltonian in the new variables takes on the following form

$$H = N_1 + N_2 + \epsilon V(z, \bar{z}, \epsilon), \tag{33}$$

where $N_1 = |z_1|^2$, $N_2 = |z_2|^2$ and $V(z, \bar{z}, \epsilon)$ is a real polynomial, i.e., has the property $V(z, \bar{z}, \epsilon) = \bar{V}(\bar{z}, z, \epsilon)$ and $\bar{V}$ is the polynomial obtained from V by complex conjugation of its coefficients. The transformation to the z variables is not quite canonical, but gives rise to Hamiltonian equations that can be written in Heisenberg form:

$$\dot{z}_k = i[H, z_k]. \tag{34}$$

Here, [,] denotes the Poisson bracket in a form that generalizes Eq. (8) in an obvious way:

$$[A, B] = \sum_{k=1}^{2}\left(\frac{\partial A}{\partial \bar{z}_k}\frac{\partial B}{\partial z_k} - \frac{\partial B}{\partial \bar{z}_k}\frac{\partial A}{\partial z_k}\right). \tag{35}$$

In the paper (Kummer 76)[4] I have shown that in the case of the semisimple 1:1 resonance the normal form of the perturbation is a formal power series in the variables

$$\mathbf{M} = \frac{1}{2} < z, \sigma z >, \quad J = \frac{1}{2} < z, z >, \tag{36}$$

where $< z, w >= \bar{z}_1 w_1 + \bar{z}_2 w_2 = z^\dagger w$ denotes the usual inner product in $\mathcal{C}^2$ and

$$\sigma = \left\{ \begin{pmatrix} 0 & 1 \\ 1 & 0 \end{pmatrix}, \begin{pmatrix} 0 & i \\ -i & 0 \end{pmatrix}, \begin{pmatrix} 1 & 0 \\ 0 & -1 \end{pmatrix} \right\} \tag{37}$$

is the vector of Pauli matrices. In place of M_1, M_2 one can also use

$$M_+ = z_1\bar{z}_2, \quad M_- = \bar{z}_1 z_2, \tag{38}$$

which are related to the previous variables through the relations:

$$M_1 = \frac{1}{2}(M_+ + M_-), \quad M_2 = \frac{1}{2i}(M_+ - M_-). \tag{39}$$

The Poisson bracket relations correspond to those of su(2)

[4]For specific examples see also, e.g., Churchill, Kummer and Rod [1983], Churchill and Lee [1984].

$$[M_+, M_-] = 2M_3, \ \ [M_3, M_\pm] = \pm M_\pm. \tag{40}$$

Alternatively, these relations can be written in the form

$$[M_1, M_2] = iM_3 \tag{41}$$

and cyclic.

The association $z_1, z_2 \to \mathbf{M} := (M_1, M_2, M_3)$, defines a map whose image of the 3 - sphere J = const. is the 2 - sphere

$$M_1^2 + M_2^2 + M_3^2 = J^2. \tag{42}$$

This map was later recognized as the Hopf map and interpreted to be the moment map of the natural SU(2) action on $\mathcal{C}^2$. Accordingly, the variables $J, \mathbf{M}$ are now called Hopf variables. (See, e.g., Cushman and Rod [1982].)

Now, by a simple reinterpretation of the symbols, the same formalism applies to the corresponding quantum mechanical system: In order to do this most clearly let us look at what we have done from an algebraic point of view. Consider now the Poisson algebra generated by five generators $1, z_1, \bar{z}_1, z_2, \bar{z}_2$, subject to the relations

$$[\bar{z}_k, z_k] = 1 \ \ (k = 1, 2) \tag{43}$$

and all other brackets zero. It is clear that all elements of the algebra generated in this way are polynomials in the four variables $z_1, \bar{z}_1, z_2, \bar{z}_2$.

Theorem 1.2 *If we assume that the algebra is commutative we recover the situation in classical mechanics. In other words the bracket must be the Poisson bracket as given above.*

On the other hand, like in the case of systems of one degree of freedom, we obtain quantum theory by simply dropping the assumption of commutativity and interpreting the bracket as commutator. This means that z_1, z_2 become creation operators and $\bar{z}_1, \bar{z}_2$ are annihilation operators. Since the normal form theorem can be formulated within the framework of a (graded) Lie algebra it still makes sense within the quantum Poisson algebra. Whereas the definition of the M_k - variables remains unchanged, we adhere to the custom in quantum mechanics to use symbols $N_1 = z_1\bar{z}_1, N_2 = z_2\bar{z}_2$ for the number operators and define $J = \frac{1}{2}(N_1 + N_2)$ as in the classical context so that the unperturbed Hamiltonian in the 1:1 resonance becomes $2J + 1$. Of course, due to the non-commutativity of the algebra, the squares of the M_k- variables now add up to $J(J+1)$ instead of J^2.

However, there is one crucial difference between normal forms for classical oscillators and quantum mechanical ones: In classical mechanics the homogeneity degree of the perturbing polynomial defines a natural grading. Terminating the normal form at a certain power will usually yield an approximation which can be used to say something (via KAM theory) about the flow induced by the original Hamiltonian close to the origin of phase space.

In order for the normal form in quantum mechanics to make sense the second grading by means of powers of the coupling constant ϵ becomes extremely important. This is because a polynomial that is homogeneous in the annihilation and creation operators loses this property if one rearranges their order by using the commutation relations $[\bar{z}_k, z_k] = 1$. On the other hand it is clear that any polynomial in which the degree of its homogeneous components has a fixed parity can be made homogeneous by use of the commutation relations. These remarks have the following consequences for the quantum normal form: In order to express a given term which is in normal form in terms of the Hopf variables it is usually necessary to rearrange the order in which annihilation and creation operators appear after the projection onto the null space $\mathcal{N}$ and in this way one introduces lower order terms. The upshot of this discussion is that in quantum mechanics rather than truncating the normal form at a definite homogeneity degree one should truncate at a definite power of the coupling constant. Still the question remains: What is the significance of the quantum normal form for the original Hamiltonian? A student of mine, Raghuramaiha Gompa, wrote a Ph.D thesis entitled "Approximations to the quantummechanical time evolution" in which he used the normal form to construct approximate solutions to the time - dependent Schrödinger equation, all with rigorous error estimates. Related work has been done before him by B. Lemlih and J.A. Ellison on so called quantum averaging (see Lemlih and Ellison [1985]).[5] The application of our scheme to anharmonic oscillators has been published in Kummer and Gompa [1988].[6] However, more should be possible. One should also be able to approximate the spectrum of the original Hamilton operator by the one of a normal form.

Conjecture: *For sufficiently small ϵ the original Hamiltonian has pure point spectrum and its spectrum is approximated asymptotically for $\epsilon \to 0$ by the point spectrum of a corresponding normal form.*

As just mentioned above, unlike in classical mechanics, even in the case in which the perturbing polynomials W_k are homogeneous polynomials of degree $k+2$ of the position variables alone (and therefore there is no ambiguity of quantization) the corresponding normalized polynomials when expressed in the Hopf variables are in general not homogeneous. Still, it sometimes happens that if in a classical normal form the substitution $J \to J + \frac{1}{2}$ is made the corresponding quantum normal form differs from the classical one by a term of order two less in the Hopf variables. For instance this is the case in the Henon - Heiles Hamiltonian [7]

$$H = \frac{1}{2}(p_1^2 + q_1^2) + \frac{1}{2}(p_2^2 + q_2^2) + \epsilon(q_1^2 q_2 + \mu q_2^3). \tag{44}$$

The dominating term in the classical normal form can be written in the form:

$$2J + \epsilon^2 K_2(J, \mathbf{M}) + \mathcal{O}(\epsilon^3), \tag{45}$$

---

[5]See also A. Ben Lemlih's Thesis: An Extension of the Method of Averaging to Partial Differential Equations, the University of New Mexico, Albuquerque, New Mexico.

[6]For an informal discussion of this approach see Kummer [1991].

[7]For a somewhat different approach to the quantummechanical Henon-Heiles Hamiltonian and an interesting theorem of the associated eigenfunctions see Uwano [1989].

where

$$K_2(J,\mathbf{M}) = \frac{1}{2}(a_1M_1^2 + a_2M_2^2 + a_3M_3^2) + J(b_1M_1 + b_2M_2 + b_3M_3) + a_0J^2 \quad (46)$$

with

$$a_0 = -\frac{1}{4}[\frac{13}{3} + 12\mu + 15\mu^2], \ \ a_1 = -a_2 = -(2-\mu), \ \ a_3 = \ \frac{1}{2}[1 + 12\mu - 15\mu^2], \quad (47)$$

$$b_1 = b_2 = 0, \ \ b_3 = \frac{15}{2}(\mu^2 - \frac{1}{9}). \quad (48)$$

It turns out that this is also exactly a quantum mechanical normal form except for a replacement $J \to J + \frac{1}{2}$ and an additional additive constant term $\frac{1}{16} - \frac{7}{16}\mu^2$. However, this is not the general situation, even if the potential is a function of the position variable only. A counter example is provided by the following Hamiltonian:

$$H = \frac{1}{2}(p_1^2 + q_1^2) + \frac{1}{2}(p_2^2 + q_2^2) + \beta_1 q_1^2 q_2 + \beta_2 q_2^2 q_1. \quad (49)$$

The dominating term in the classical normal form can be written as in Eq. (46) with

$$a_0 = -\frac{13}{12}(\beta_1^2 + \beta_2^2), \ a_1 = -a_2 = -2(\beta_1^2 + \beta_2^2), \ a_3 = \frac{1}{2}(\beta_1^2 + \beta_2^2), \quad (50)$$

$$b_1 = -\frac{10}{3}\beta_1\beta_2, \ \ b_2 = 0, \ \ b_3 = \frac{5}{6}(\beta_2^2 - \beta_1^2). \quad (51)$$

This time the switch to the quantum normal form requires not only the replacement $J \to J + \frac{1}{2}$ and the correction by a constant additive term, namely $\frac{1}{16}(\beta_1^2 + \beta_2^2)$, but also the correction by the additive term $\frac{3}{2}\beta_1\beta_2 M_1$ which is linear in the Hopf variables. However, the following relationship between classical and quantum normal form seems obvious:

Conjecture: *Suppose that the perturbation is a polynomial in the position variables only. If in a quantum normal form* $2J+1+\sum_{m=1}\epsilon^m K_q^{(m)}(J,\mathbf{M})$, $2J+1$ *is replaced by* $2J$ *and in each* $K_q^{(m)}(J,\mathbf{M})$ *only the terms of highest homogeneity degree are retained (where now products of the Hopf variables are allowed to commute) then a classical normal form* $2J + \sum_{m=1}\epsilon^m K_c^{(m)}(J,\mathbf{M})$ *is obtained.*

How about the motion of the vector $\mathbf{M}$ in the course of time? Using the Poisson structure and the bracket relations one easily proves in the classical setting that the motion of the vector $\mathbf{M}$ is governed by Euler - equations of the form

$$\dot{\mathbf{M}} = \nabla_{\mathbf{M}} K \times \mathbf{M}, \quad (52)$$

where K is the (truncated) normal form. In order to obtain a neat quantum mechanical interpretation of the Euler equation we best use Bargmann's representation of the commutation relations on his Hilbert space $\mathcal{F}^2$ of analytic functions in the variables z_1 and z_2. In this interpretation z_1 is the multiplication operator that replaces $\psi \in \mathcal{F}^2$ by $z_1\psi$ and $\bar{z}_1$ symbolizes the partial derivative with respect to z_1 with analogous interpretations for z_2 and $\bar{z}_2$. The M_k's now completely agree with those used by Bargmann [1962] in his discussion of the representation of the rotation group. The Euler equations become a countable set of finite linear matrix equations as follows: The Bargmann Hilbert - space decomposes into a direct sum of finite dimensional spaces $\mathcal{F}^2 = \sum \oplus \mathcal{P}_{2j}$, where j ranges over all non - negative half-integers, $\mathcal{P}_{2j}$ being the space of homogeneous polynomials of degree 2j. Each space is invariant under the action of the M_k's and therefore also under the flow of the quantum mechanical normal form Hamiltonian thus leading to a linear matrix differential equation in each subspace $\mathcal{P}_{2j}$.

2 Conversion of the perturbed planar Kepler problem into an oscillator problem.

Given a Hamiltonian system

$$\frac{dx}{dt} = \frac{\partial H}{\partial p}, \quad \frac{dp}{dt} = -\frac{\partial H}{\partial x} \tag{53}$$

and a positive (continuous) function $f(x,p)$, we consider the associated Hamiltonian system

$$\frac{dx}{ds} = \frac{\partial K}{\partial p}, \quad \frac{dp}{ds} = -\frac{\partial K}{\partial x}, \tag{54}$$

with Hamiltonian $K := f(x,p)(H-C)+k$, (C, k suitable constants) and new time $s = s(t)$. Here, $s(t)$ is obtained by inverting the monotone (C^1-) function

$$t(s) = \int_0^s f(x(v), p(v))dv. \tag{55}$$

We claim that if $x(s), p(s)$ is a solution of K on the on the surface $K = k \Leftrightarrow H = C$ then $x(s(t)), p(s(t))$ is such a solution of H. Indeed, $\frac{dx}{dt} = \frac{\frac{dx}{ds}}{\frac{dt}{ds}} = \frac{\partial H}{\partial p} + \frac{1}{f}\frac{\partial f}{\partial p}(H-C)$ and a similar equation for $\frac{dp}{dt}$. Using the foregoing construction the planar Kepler problem for negative energies can be converted into the $1:1$ resonance and the theory pertaining to this resonance can be applied to the present problem. In classical mechanics this conversion is manufactured as follows: We identify the configuration plane with the complex x - plane and the momentum plane with the complex p - plane. The Hamiltonian of the perturbed planar Kepler problem is

$$H = \frac{1}{2}p\bar{p} - \frac{k}{|x|} + \epsilon W(x,p,\epsilon). \tag{56}$$

Setting $|x| = r$ we assume that $W(x,p,\epsilon)$ is a polynomial of the form:[8]

$$W(x,p,\epsilon) := W_0(x, rp, \epsilon) + Re(pW_1(x, rp, \epsilon)). \tag{57}$$

The equation of motion induced by H are

[8]This assumption on W will guarantee that the derived functions $\tilde{W}, V$ (see (62) and (65)) are again real valued polynomials of the indicated variables and their conjugate complex counterparts.

$$\frac{dx}{dt} = 2\frac{\partial H}{\partial \bar{p}} = p, \quad \frac{dp}{dt} = -2\frac{\partial H}{\partial \bar{x}}. \tag{58}$$

They correspond to the one-form $\Theta = Re\bar{p}dx$. We introduce the fictitious time s by means of the recipe

$$\frac{dt}{ds} = |x| := r, \tag{59}$$

i.e., we set $f = |x|$ in the preceding construction leading to the modified Hamiltonian

$$K = r(H + \frac{1}{2}\omega^2) + k = \frac{r}{2}(p\bar{p} + \omega^2) + \epsilon r W(x, p, \epsilon). \tag{60}$$

Then, as explained above, reparametrizing any trajectory of the Hamiltonian K on the surface $K = k \Leftrightarrow H = -\frac{\omega^2}{2}$ by t yields a trajectory of H. Now we subject K to the Levi Civita transformation

$$x = q^2, \quad p = \frac{1}{\bar{q}}w \tag{61}$$

to obtain

$$K = \frac{1}{2}|q|^2(\frac{|w|^2}{|q|^2} + \omega^2) + \epsilon|q|^2 W(q^2, \frac{1}{\bar{q}}w, \epsilon) = \frac{1}{2}(|w|^2 + \omega^2|q|^2) + \epsilon\tilde{W}(q, w, \epsilon). \tag{62}$$

The classical equations of motion in the new variables are:

$$\dot{q} = \frac{\partial K}{\partial \bar{w}}, \quad \dot{w} = -\frac{\partial K}{\partial \bar{q}} \tag{63}$$

since they correspond to the 1-form: $\Theta = 2Re\bar{w}dq$. A further transformation:

$$q = \frac{1}{\sqrt{2\omega}}(\bar{z_2} + z_1), \quad w = \frac{1}{i}\sqrt{\frac{\omega}{2}}(\bar{z_2} - z_1) \tag{64}$$

yields

$$K = \frac{\omega}{2}(N_1 + N_2) + \epsilon V(z_1, z_2, \epsilon) = \omega J + \epsilon V(z_1, z_2, \epsilon), \tag{65}$$

where $N_1 = z_1\bar{z_1}, N_2 = z_2\bar{z_2}, J := \frac{1}{2}(N_1 + N_2)$. Since also

$$\Theta = \frac{1}{i}(\bar{z}_1 dz_1 + \bar{z}_2 dz_2), \tag{66}$$

this means of course that the perturbed Kepler problem for negative energies has been replaced by a perturbation problem of two oscillators in 1 : 1 resonance.[9]

[9]Note however, that (z_1, z_2) and $(-z_1, -z_2)$ represent the same state as a consequence of the Levi- Civita transformation.

For later reference we also note:

$$\Theta \cong N_1 d\alpha_1 + N_2 d\alpha_2 = Jdu + M_3 d\gamma, \tag{67}$$

where $\alpha_k = arg z_k$ $(k = 1, 2)$, $\gamma = \alpha_1 - \alpha_2$, $u = \alpha_1 + \alpha_2$.

The Kepler - flow has been trivialized in the sense that it has been replaced by the circular flow:

$$z_1 = z_1^{(0)} exp(i\frac{\omega}{2}s), \quad z_2 = z_2^{(0)} exp(i\frac{\omega}{2}s). \tag{68}$$

Introducing the generators of one - parameter subgroups of $Sp(2, \mathcal{R})$:

$$Q_+ := \frac{1}{2}(\bar{z}_2{}^2 + z_1^2), \quad M_+ = z_1 \bar{z}_2, \tag{69}$$

the position variable x becomes

$$x = \frac{1}{\omega}(Q_+ + M_+). \tag{70}$$

More generally, a potential function V of our class becomes a polynomial in the generators of one-parameter subgroups of $Sp(2, \mathcal{R})$. A convenient choice of the ten generators of this group is [10]

$$\begin{array}{lll} U(2) = \text{max. compact subgroup} & & Re\ Im \\ \quad J := \frac{1}{2}(N_1 + N_2) & M_3 = \frac{1}{2}(N_1 - N_2) & z_1\bar{z}_2 \\ & & \bar{z}_2{}^2 + z_1^2 \\ & & z_1 z_2 \\ & & \bar{z}_2{}^2 - z_1^2 \end{array} \tag{71}$$

The normal form theorem implies that any perturbation by means of a canonical transformation can be brought into the form

$$\omega J + \epsilon K_c(J, \mathbf{M}, \omega, \epsilon), \tag{72}$$

where $\mathbf{M}, J$ listed in the first line of the table (71) generate the subgroup $U(2)$. $M_+ = M_1 + iM_2$ represents the Laplace (= negative of Lenz) vector, M_3 the angular momentum and each point of the sphere (42) (see Figure 3) corresponds to an oriented Kepler orbit. In particular the equator corresponds to the collision orbits and the poles to direct and retrograde circular solution. Indeed,

$$Im(\bar{x}p) = x_1 p_2 - x_2 p_1 = Im\bar{q}w = \frac{1}{2}(N_1 - N_2) = M_3. \tag{73}$$

Also, expressing x in terms of the variables introduced in connection with Eq. (67), we find

$$x = q^2 = \frac{1}{2\omega}(\bar{z}_2^2 + z_1^2 + 2z_1\bar{z}_2) = \frac{1}{2\omega}(N_2 e^{-2i\alpha_2} + N_1 e^{2i\alpha_1} + 2\sqrt{N_1 N_2} e^{i\gamma}) \tag{74}$$

$$= \frac{1}{2\omega} e^{i\gamma}(N_2 e^{-iu} + N_1 e^{iu} + 2\sqrt{N_1 N_2}) = \frac{1}{\omega} e^{i\gamma}[J \cos u + iM_3 \sin u + \sqrt{J^2 - M_3^2}]. \tag{75}$$

---

[10]For another choice see Baumgarte [1978].

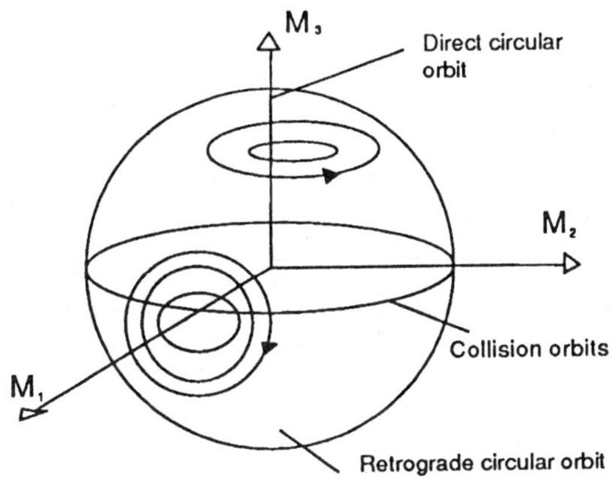

Figure 3 Sphere whose points represent oriented Kepler ellipses in 2- dim. problem with sample flowlines induced by a normalized perturbation.

In case of the pure Kepler problem (with k=1) $\omega = \frac{1}{J}$ and $u = \omega s$ so that modulo the rotation through the angle γ :

$$x_1 = a(\cos u + e), \quad x_2 = b \sin u, \tag{76}$$

where $a = J^2$, $b = JM_3$, $e = \frac{1}{J}\sqrt{J^2 - M_3^2}$. Observe that setting $c = ae = J\sqrt{J^2 - M_3^2}$, we have $b^2 + c^2 = a^2$.
Kepler equation:

$$\frac{dt}{ds} = |x| = |q|^2 = \frac{1}{2\omega}(N_1 + N_2 + 2Re(z_1 z_2)) = J(J + \sqrt{J^2 - M_3^2}\cos u) \tag{77}$$

$$= a(1 + e\cos u) \Rightarrow \frac{dt}{du} = \frac{a}{\omega}(1 + e\cos u) = a^{3/2}(1 + e\cos u) \tag{78}$$

$$\Rightarrow t = a^{3/2}(u + e\sin u) := a^{3/2}\ell, \tag{79}$$

ℓ =mean anomaly.
In order to obtain the physical motion from the flow on the sphere in the perturbed problem we use the symplectic coordinates J, u, M_3, γ. Since $\gamma(s) = argM_+(s)$ the motion of the last two variables is determined from the flow on the sphere. In oder to obtain $u(s)$ we integrate the equation $\dot{u} = \frac{\partial K}{\partial J}$, where we have set $M_+ = \sqrt{J^2 - M_3^2}e^{i\gamma}$. ω is determined from the fact that $K = 1$. Finally, the physical time is obtained as a function of the fictitious time s by integrating Equation (77). In Kummer [1979] these methods are applied to the planar restricted three body problem.

Much of the foregoing development can be carried over to a quantum mechanical setting, except that the variables x, p are non-commutative, but satisfy the commutation relations

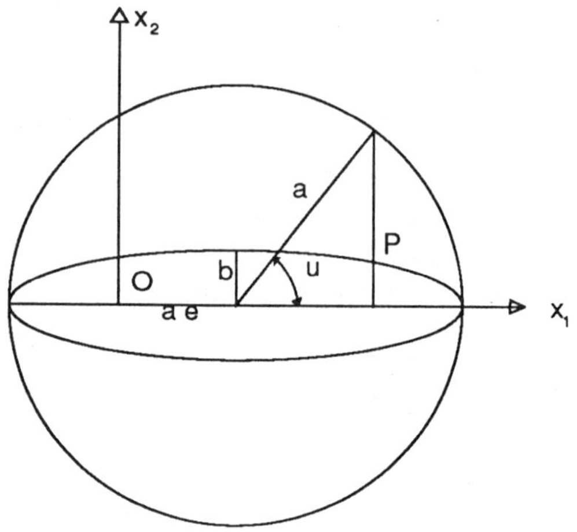

Figure 4 Elliptical orbit.

$$[\bar{p}, x] = [p, \bar{x}] = \frac{2}{i}. \tag{80}$$

Imposing on the variables w, q the condition

$$[\bar{w}, q] = [w, \bar{q}] = \frac{1}{i}, \tag{81}$$

the commutation relations for x, p are a consequence of the Levi - Civita transformation

$$x = q^2, \quad p = \frac{1}{\bar{q}} w. \tag{82}$$

Moreover, the variables z_1, z_2 and $\bar{z}_1, \bar{z}_2$ introduced earlier through the relations (64) are creation and annihilation operators, i.e. they satisfy

$$[\bar{z}_1, z_1] = 1, \quad [\bar{z}_2, z_2] = 1. \tag{83}$$

Again this transformation converts the original perturbed Kepler problem into a perturbation problem of two harmonic oscillators in 1:1 resonance: Indeed, since

$$K = r(H + \frac{1}{2}\omega^2) + k = \omega(J + \frac{1}{2}) + \epsilon\tilde{V}(z_1, z_2, \epsilon), \tag{84}$$

the Schrödinger equation

$$H\psi = -\frac{1}{2}\omega^2\psi \tag{85}$$

becomes $[\omega(J+\frac{1}{2})+\epsilon\tilde{V}(z_1,z_2,\epsilon)]\psi = k\psi$. Bringing the perturbation into quantum normal form $K_q(J,\mathbf{M},\omega,\epsilon)$ the transformed Schrödinger equation is

$$[\omega(J+\frac{1}{2})+\epsilon K_q(J,\mathbf{M},\omega,\epsilon)]\psi = k\psi. \tag{86}$$

Since the transformation to normal form is formally unitary one hopes that one obtains an approximation to the correct spectrum of the original problem. In this fashion the two dimensional perturbed Kepler problem has been subsumed under the perturbed 1 : 1 resonance in classical and quantum mechanics. For a specific application of these ideas see Kummer [1991].

3 Results in the three dimensional problem.

(This section reviews and builds on Kummer [1982], [1983], [1984], [1985]). For background material and related work see Baumgarte [1978], Kustaanheimo and Stiefel [1965], Stiefel and Scheifele [1971], Milnor [1983].)

Our goal is to convert the (perturbed) Kepler problem into a problem of (perturbations) of four harmonic oscillators. This is done by means of a computational technique, called σ- gymnastics, to which we now turn: Clearly, the matrix $\frac{1}{\sqrt{2}}S$ with $S = \begin{pmatrix} 1 & 0 & 0 & 1 \\ 0 & 1 & i & 0 \\ 0 & 1 & -i & 0 \\ 1 & 0 & 0 & -1 \end{pmatrix}$ is unitary, so that $SS^\dagger = 2\mathcal{I}, S^\dagger S = 2\mathcal{I}$, where $\mathcal{I}$ is the 4 by 4 unit matrix. This implies that the four Pauli matrices $\sigma^\nu = (\sigma^0,\sigma)$ $(\nu = 0,1,2,3; \sigma^0 =$ the 2 by 2 unit matrix) satisfy

$$\sum_{\nu=0}^{3} \sigma^\nu{}_{\alpha\beta}\sigma^\nu{}_{\delta\gamma} = 2\delta_{\alpha\gamma}\delta_{\beta\delta}, \quad tr(\sigma^\nu\sigma^\mu) = 2\delta^{\nu\mu}. \tag{87}$$

Multiplying the first by $\bar{u}_\alpha, v_\beta, \bar{w}_\delta, z_\gamma$ and enforcing the Einstein summation convention we obtain

$$\sum_{\nu=0}^{3} < u,\sigma^\nu v >< w,\sigma^\nu z >= 2 < u,z >< w,v > . \tag{88}$$

Because σ^ν is Hermitian: $\overline{< w,\sigma^\nu z >} =< z,\sigma^\nu w >$. Hence,

$$\sum_{\nu=0}^{3} < u,\sigma^\nu v > Im < w,\sigma^\nu z >= \frac{1}{i}[< u,z >< w,v > - < u,w >< z,v >]. \tag{89}$$

Applying this relation to the Kustaanheimo-Stiefel map $\pi : \mathcal{C}^2\backslash\{0\}\times\mathcal{C}^2 \to \mathcal{R}^3\backslash\{0\}\times\mathcal{R}^3$ defined by means of the formulae:

$$\mathbf{x} =< z,\sigma z >, \quad \mathbf{p} =< z,z >^{-1} Im < w,\sigma z >, \tag{90}$$

we obtain first by setting $u = w, v = z$ and forming the imaginary part in Equation (89):

$$\mathbf{p}\cdot\mathbf{p} < z.z >^2 + (Im < w,z >)^2 = -Re(< w,z >)^2 + < w,w >< z,z >, \tag{91}$$

i.e.,

$$\mathbf{p}\cdot\mathbf{p} < z.z >^2 = -(Re < w,z >)^2 + < w,w >< z,z >, \tag{92}$$

and by setting $u = z$ and $v = dz$ and subsequently forming the real part in that relation we obtain:

$$< z,z > \frac{1}{2}\mathbf{p}\cdot d\mathbf{x} + Im < w,z > Re < z,dz >=$$

$$Im[< z,z >< w,dz > - < z,w >< z,dz >],$$

or

$$< z,z > \frac{1}{2}\mathbf{p}\cdot d\mathbf{x} =< z,z > Im < w,dz > -Re < w,z > Im < z,dz > . \tag{93}$$

The last relation has the interpretation $\pi^*\Theta_0 = i^*\Theta$, where $\Theta_0 = \mathbf{p}\cdot d\mathbf{x}$, $\Theta = 2Im < w,dz >$ are the canonical one-forms on $\mathcal{R}^3\backslash\{0\}\times\mathcal{R}^3, \mathcal{C}^2\backslash\{0\}\times\mathcal{C}^2$, respectively and $i : I^{-1}(0)\to\mathcal{C}^2\backslash\{0\}\times\mathcal{C}^2$ is the injection of the 7-dimensional cone $I := Re < w,z >= 0$ into $\mathcal{C}^2\backslash\{0\}\times\mathcal{C}^2$. Moreover, the physical states can be identified with circles that are flow lines of the Hamiltonian $I := Re < w,z >$ on this cone. In other words, the phase space $\mathcal{R}^3\backslash\{0\}\times\mathcal{R}^3$ of the Kepler problem is represented by the reduced space [11] $I^{-1}(0)/U(1)$ so that we can simply set

$$\mathbf{p}\cdot d\mathbf{x} = 2Im < w,dz >,\ \ \mathbf{p}\cdot\mathbf{p} =< z,z >^{-1}< w,w >, r = |x| =< z,z > . \tag{94}$$

Applying this to the "preregularized" perturbed Kepler Hamiltonian

$$K = K_0 + \epsilon r W(\mathbf{x},\mathbf{p},\epsilon),\ \ K_0 := \frac{1}{2}r(\mathbf{p}^2+\omega^2) \tag{95}$$

we obtain

$$K = K_0 + \epsilon\tilde{V}(z,w,\epsilon),\ \ K_0 = \frac{1}{2}[< w,w > +\omega^2 < z,z >]. \tag{96}$$

Since we make the same assumptions about the polynomial W [12] as in the two-dimensional case $\tilde{V}$ is again a polynomial of the indicated variables. As explained earlier in connection with the two-dimensional problem the Hamiltonian K induces on the energy surface $K = k \Leftrightarrow H = -\frac{\omega^2}{2}$ the same flow lines as the original perturbed Kepler Hamiltonian

$$H = \frac{\mathbf{p}^2}{2} - \frac{k}{r} + \epsilon W(\mathbf{x},\mathbf{p},\epsilon), \tag{97}$$

except that they are reparametrized by the fictious time s instead of physical time t, where the two times are related through $\frac{dt}{ds} = r$. Subjecting K to a further transformation:

$$\eta = \frac{1}{\sqrt{2}}(\sqrt{\omega}z + \frac{1}{\sqrt{\omega}}w),\ \ \zeta = \frac{1}{\sqrt{2}}(\sqrt{\omega}z - \frac{1}{\sqrt{\omega}}w) \tag{98}$$

[11] For the theory of the reduced space see, e.g., Marsden and Weinstein [1974], Kirillov [1976], Abraham and Marsden [1978], Moser [1980].

[12] with obvious modifications due to the change in dimensionality.

with inverse

$$z = \frac{1}{\sqrt{2\omega}}(\eta + \zeta), w = \sqrt{\frac{\omega}{2}}(\eta - \zeta)$$

leads to the desired result

$$K_0 = \frac{1}{2}\omega[< \eta, \eta > + < \zeta, \zeta >] := 2J\omega, \tag{99}$$

$$\theta := 2Im < w, dz > \simeq Im[< \eta, d\eta > - < \zeta, d\zeta >] \tag{100}$$

$$\simeq \frac{1}{i}(\eta^\dagger, \zeta^\dagger)\mathcal{J}\begin{pmatrix} d\eta \\ d\zeta \end{pmatrix}, \text{ where } \mathcal{J} = \begin{pmatrix} \sigma_0 & 0 \\ 0 & -\sigma_0 \end{pmatrix}, \tag{101}$$

namely the Kepler Hamiltonian is converted into one describing four harmonic oscillators in resonance, as predicted. However, K_0 as well as the perturbation has to be restricted to the cone $I = \frac{1}{2}[< \eta, \eta > - < \zeta, \zeta >] = 0$ [13] on which the physical states are circles $s \to (\eta e^{is}, \zeta e^{is})$. On the other hand the Kepler flow is represented by the circular flow $s \to (\eta e^{i\frac{\omega}{2}s}, \zeta e^{-i\frac{\omega}{2}s})$. Symplectic transfomations leaving the constraint invariant belong to the group $U(2,2)$, i.e., they are matrices $\mathcal{U}$ satisfying $\mathcal{U}^\dagger \mathcal{J} \mathcal{U} = \mathcal{J}$. I will identify the corresponding Lie algebra $u(2,2)$ and its dual $u(2,2)^*$ with the set of matrices characterized by the property that $\mathcal{J}A$ is Hermitian. Correspondingly, the Lie bracket differs from the commutator by a factor $\frac{1}{i}$, i.e.,

$$[\mathcal{A}, \mathcal{B}] := \frac{1}{i}(\mathcal{AB} - \mathcal{BA}), \tag{102}$$

where $\mathcal{A}, \mathcal{B} \in u(2,2)$. The pairing between $u(2,2)$ and $u(2,2)^*$ is given by the expression $< \mathcal{B}, \mathcal{A} >= tr(\mathcal{J}\mathcal{B}\mathcal{J}\mathcal{A})$ $(\mathcal{B} \in u(2,2)^*, \mathcal{A} \in u(2,2))$. Moreover adjoint and coadjoint actions are described by the assignments $\mathcal{A} \to \mathcal{U}\mathcal{A}\mathcal{U}^{-1}$, $\mathcal{B} \to (\mathcal{U}^\dagger)^{-1}\mathcal{B}\mathcal{U}^\dagger$. Since the one-parameter group corresponding to $\mathcal{A}$ induces the flow $exp(i\mathcal{A}s)\begin{pmatrix} \eta \\ \zeta \end{pmatrix}$ the associated fundamental vector field is $X^{\mathcal{A}}_{\eta,\zeta} = i\mathcal{A}\begin{pmatrix} \eta \\ \zeta \end{pmatrix}$. Since the action of $u(2,2)$ is exact symplectic the flow is induced by the Hamiltonian: [14]

$$\Psi_{\mathcal{A}}(\eta, \zeta) := \Theta(X^{\mathcal{A}}_{\eta,\zeta}) = (\eta^\dagger, \zeta^\dagger)\mathcal{J}A\begin{pmatrix} \eta \\ \zeta \end{pmatrix} =$$

$$tr(\mathcal{J}\Psi(\eta, \zeta))\mathcal{J}\mathcal{A}) =< \Psi(\eta, \zeta), \mathcal{A} >$$

where $\Psi(\eta, \zeta) : \mathcal{C}^4 \backslash \{0\} \to u(2,2)^*$ is the moment map

$$\Psi(\eta, \zeta) = \begin{pmatrix} \eta\eta^\dagger & \eta\zeta^\dagger \\ -\zeta\eta^\dagger & -\zeta\zeta^\dagger \end{pmatrix}. \tag{103}$$

The image of the cone $I^{-1}(0)$ under Ψ consists of traceless matrices whose associated Hermitian matrices have rank one. The infinitesimal equivariance of the moment map finds its expression in the formula $\{\Psi_{\mathcal{A}}, \Psi_{\mathcal{B}}\} = -i\Psi_{[\mathcal{A},\mathcal{B}]}$, where $\{,\}$

---

[13] In the last five references of this section a generalization of the Kepler problem known as conformal Kepler problem is considered. Its reduction w.r. to $U(1)$ leads to hyperboloids $I = c \neq 0$.

[14] For the theory behind this construction see Abraham and Marsden [1978].

signifies the Poisson bracket w.r. to the form $id\theta$ and $[,]$ is the Lie bracket in $u(2,2)$. Important elements of $u(2,2)$ which correspond to physically meaningful quantities are:

$$\mathcal{M} = \frac{1}{2}\begin{pmatrix} \sigma & 0 \\ 0 & 0 \end{pmatrix}, \mathcal{N} = \frac{1}{2}\begin{pmatrix} 0 & 0 \\ 0 & \sigma \end{pmatrix}, \mathcal{J}, \tag{104}$$

$$\mathcal{Q} + i\mathcal{P} = [\frac{1}{i}\begin{pmatrix} 0 & \sigma^0 \\ 0 & 0 \end{pmatrix}, \begin{pmatrix} 0 & \sigma \\ 0 & 0 \end{pmatrix}]. \tag{105}$$

The associated Hamiltonians are

$$\mathbf{M} =< \Psi(\eta,\zeta), \mathcal{M} >= \frac{1}{2} < \eta, \sigma\eta >, \tag{106}$$

$$\mathbf{N} =< \Psi(\eta,\zeta), \mathcal{N} >= -\frac{1}{2} < \zeta, \sigma\zeta >, \ J = \frac{1}{4} < \Psi(\eta,\zeta), \mathcal{J} > \tag{107}$$

and

$$Q + iP =< \Psi(\eta,\zeta), \mathcal{Q} + i\mathcal{P} >= (-i < \eta, \zeta >, < \eta, \sigma\zeta >). \tag{108}$$

In particular,

$$P_0 = -Re < \eta, \zeta >, \ \mathbf{Q} = Re < \eta, \sigma\zeta >, \ \mathbf{P} = Im < \eta, \sigma\zeta > . \tag{109}$$

The phase coordinates find the following expressions in terms of these Hamiltoninans

$$\mathbf{x} = \frac{1}{\omega}(\mathbf{R} + \mathbf{Q}), \ \ r\mathbf{p} = \mathbf{P}, \ \ r =< z, z >= \frac{1}{\omega}(2J - P_0). \tag{110}$$

Here, $\mathbf{R} := \mathbf{M} - \mathbf{N}$ has the physical meaning of the Laplace vector. Also, $\mathbf{L} := \mathbf{M} + \mathbf{N}$ is the angular momentum. Notice that $\mathbf{M}, \mathbf{N}$ are the generators of the maximal compact $SU(2) \times SU(2)$ subgroup of $SU(2,2)$. This is precisely the subgroup of $SU(2,2)$ that leaves the Kepler Hamiltonian K_0 invariant. Accordingly, normal form theory implies that the total Hamiltonian appears as a formal power series over the Lie algebra $u(1) \times su(2) \times su(2)$, with corresponding " moment map" $J, \mathbf{M}, \mathbf{N}$:

$$2J\omega + \epsilon K(J, \mathbf{M}, \mathbf{N}, \omega, \epsilon), \tag{111}$$

where $\mathbf{M}, \mathbf{N}$, defined in Equation (106) and Equation (107), satisfy $\mathbf{M}^2 = J^2$, $\mathbf{N}^2 = J^2$. Thus the reduction with respect to the torus action $U(1) \times U(1)$ defined by the gauge group and the Kepler flow leads to the well known reduced space $S^2 \times S^2$ whose points correspond to oriented Kepler ellipses. Since, $\mathbf{M} = \frac{1}{2}(\mathbf{L} + \mathbf{R})$, $\mathbf{N} = \frac{1}{2}(\mathbf{L} - \mathbf{R})$ the normal form can equally well be considered to be a power series in $\mathbf{L}$ and $\mathbf{R}$. The flow induced in the elements of the Kepler orbits is governed by the coupled Euler equations:

$$\dot{\mathbf{M}} = \nabla_{\mathbf{M}} K \times \mathbf{M}, \ \ \dot{\mathbf{N}} = \nabla_{\mathbf{N}} K \times \mathbf{N}. \tag{112}$$

A similar result holds for positive energies: In this case the normal form reads as follows:

$$\omega P_0 + \epsilon K(P_0, \mathbf{L}, \mathbf{Q}, \omega, \epsilon) \tag{113}$$

The orbit manifold is reprensented by the equations: $\mathbf{Q}^2 - \mathbf{L}^2 = P_0^2, \mathbf{Q} \cdot \mathbf{L} = 0$. It is easily seen to have topological character $S^2 \times \mathcal{R}^2$. The equation of motions are

$$\dot{\mathbf{Q}} = -\nabla_{\mathbf{Q}} K \times \mathbf{L} + \nabla_{\mathbf{L}} K \times \mathbf{Q}, \;\; \dot{\mathbf{L}} = \nabla_{\mathbf{Q}} K \times \mathbf{Q} + \nabla_{\mathbf{L}} K \times \mathbf{L}. \tag{114}$$

Alternatively, these equations can be written in the form

$$\dot{\mathbf{C}} = i\nabla_{\mathbf{C}} K \times \mathbf{C}, \;\; \dot{\mathbf{C}}^* = -i\nabla_{\mathbf{C}^*} K \times \mathbf{C}^*, \tag{115}$$

where $\mathbf{C}$ is the complex vector $\mathbf{C} = \frac{1}{2}(\mathbf{Q} + i\mathbf{L})$ and $\mathbf{C}^*$ its conjugate. They lie on the complex sphere: $\mathbf{C}^2 = \frac{P_0{}^2}{4}$.

4 The quantum mechanical Kepler problem

(For background material and related work see Pauli [1929], Fock [1935], Bargmann [1936], Bacry [1966], Gyoergyi [1968], Jost, and Vivarelli [1985]).

The point of departure is the Schrödinger equation

$$H\psi = -\frac{\omega^2}{2}\psi. \tag{116}$$

$\mathbf{x}$, $\mathbf{p}$ are represented via the KS transformation as in classical mechanics (see 90), except that $w, \bar{w}, z$ and $\bar{z}$ are subjected to the following commutation relations: $[w_\alpha, \bar{z}_\beta] = -\delta_{\alpha\beta}$ and $[\bar{w}_\alpha, z_\beta] = \delta_{\alpha\beta}$. Indeed, one checks that the canonical quantization condition

$$[p_k, x_\ell] = \frac{1}{i}\delta_{k,\ell} \tag{117}$$

is satisfied. Physical states are represented by functions ψ of the variables $z, \bar{z}$ that are invariant under the flow induced by I. This means that they must satisfy $\psi(e^{i\phi}z, e^{-i\phi}\bar{z}) = \psi(z, \bar{z})$. On these functions the expressions for r and $\mathbf{p} \cdot \mathbf{p}$ are identical with the classical expressions (see Equation (94)). Subjecting operators pairs $w, \bar{w}, z$ and $\bar{z}$ to the same transformation (98) as in the classical case we obtain the commutation relations

$$[\bar{\eta}_\alpha, \eta_\beta] = \delta_{\alpha\beta}, \;\; [\bar{\zeta}_\alpha, \zeta_\beta] = -\delta_{\alpha\beta}, \tag{118}$$

i.e. $\bar{\eta}$ and ζ are annihilation operators. The corresponding number operators are $Y = \eta \cdot \bar{\eta}$ and $Z = \bar{\zeta} \cdot \zeta$. In terms of these number operators we can write

$$K_0 = \frac{1}{2}\omega[Y + Z + 2] := \omega(2J + 1), \tag{119}$$

where

$$J = \frac{1}{4}(\eta \cdot \bar{\eta} + \bar{\zeta} \cdot \zeta). \tag{120}$$

In Bargmann's representation these operators act on "gauge invariant functions", i.e. functions in the kernel of $Y - Z$. Here, we think of $\bar{\eta}$ being represented by $\frac{\partial}{\partial\eta}$ and ζ by $\partial/\partial\bar{\zeta}$. More precisely, the Hilbert space of states is the subspace of the Bargmann space $\mathcal{F}^4$ of functions holomorphic in η and $\bar{\zeta}$, which corresponds to the zero sector of the integral $Y - Z$. Again the perturbation is a polynomial in the generators of one - parameter subgroups of $SU(2,2)$, the moment map having the same form as in classical mechanics except that it takes its value in $su^*(2,2) \times \mathcal{L}$, where $\mathcal{L}$ is the space of linear operators over $\mathcal{F}^4$. Correspondingly, in the formula $\{\Psi_{\mathcal{A}}, \Psi_{\mathcal{B}}\} = -i\Psi_{[A,B]}$ expressing the equivariance of the moment map, the Poisson bracket $\{\ ,\}$ is replaced by the commutator. Again the normal form only depends on the generators $J, \mathbf{M}, \mathbf{N}$ of $U(1) \times SU(2) \times SU(2)$ which however now satisfy

$$\mathbf{M}^2 = J(J+1), \quad \mathbf{N}^2 = J(J+1). \tag{121}$$

Alternatively, the normal form can be viewed as a formal function in $\mathbf{R} = \mathbf{M}-\mathbf{N}$ and $\mathbf{L} = \mathbf{M} + \mathbf{N}$. Basic Question: What can one say about the relationships between normal forms in a classical problem and the corresponding quantum mechanical problem? Rather than treating this question in full generality we restrict ourselves to physically interesting examples. Here, I assume that all quantities are dimensionless. In q. m. this can always be achieved by replacing the Hamiltonian $\tilde{H}(\mathbf{x}, \mathbf{p})$ by

$H(\mathbf{x}, \mathbf{p}) := \frac{a_0}{e^2}\tilde{H}(a_0\mathbf{x}, \frac{\hbar}{a_0}\mathbf{p})$, where $a_0 = \frac{\hbar^2}{e^2\mu}$ is the Bohr radius.[15]

For example the unperturbed Hamiltonian $H_0 = \frac{\mathbf{p}^2}{2\mu} - \frac{Ze^2}{r}$ is converted into

$$H_0 = \frac{\mathbf{p}^2}{2} - \frac{Z}{r}. \tag{122}$$

5 Examples of perturbations

For applications of normalization techniques based on the regularization in Moser [1970] to perturbations of the classical Kepler problem see Cushman [1992]. For application of Lie algebraic techniques to perturbations of the hydrogen atom see McIntosh [1971]- Wulfram [1971], Barut [1972], Englefeld [1972], Becker and Bleuler [1976], Barut and Raczka [1977].

(i) **Quadratic Zeeman - effect**:

$$\alpha V(\mathbf{x}, \mathbf{p}, \alpha) = \alpha L_3 + \frac{\alpha^2}{2}(x_1^2 + x_2^2),$$

$L_3 = x_1p_2 - x_2p_1 =$ third component of angular momentum, $\alpha = B\frac{a_0}{e^2}b_0$, where $b_0 = \frac{eh}{2c\mu} =$ Bohr magneton, $B =$ magnetic field.

(ii) **Stark - effect**:

$$V(\mathbf{x}, \mathbf{p}) = \beta x_1, \quad \beta = \mathcal{E}e\frac{a_0^2}{e^2},$$

$\mathcal{E} =$ electric field.

(iii) **An orthogonal superposition of Stark effect and Zeeman effect**:

$$V(\mathbf{x}, \mathbf{p}) = -\alpha L_3 + \beta x_1.$$

---

[15] $\hbar$ signifies Planck's constant divided by 2π.

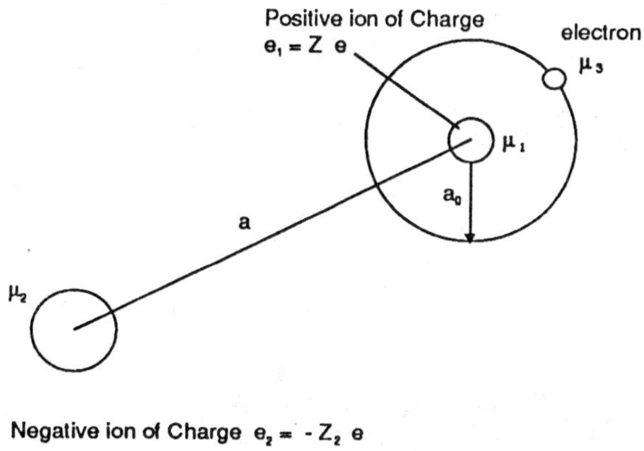

Figure 5 The Coulomb restricted three Body problem.

This case was treated classically by Deprit [1984] (1st order perturbation) and van der Meer and Cushman [1987] (2nd order perturbation). Classically, the Hamiltonian describes orbiting dust (about the earth) under radiation pressure (from the sun). In this interpretation β is proportional to the pressure, whereas γ is proportional to the rotation frequency of the earth about the sun. Although (ii) is treated quantum mechanically up to second order in the literature, I am not aware of such a treatment of (iii).

(iv) **Coulomb restricted three body problem (lunar problem)**:

$$\epsilon V(\mathbf{x},\mathbf{p},\epsilon) = -\gamma\epsilon^{3/2}L_3 - \kappa\epsilon^2 x_1 + \epsilon^3(Z^2/2)(3x_1^2 - r^2) + \mathcal{O}(\epsilon^4),$$

where

$$\gamma^2 = \frac{Z_1 Z_2 \mu_3}{\mu_{12}}, \ \kappa = Z_2(\frac{Z_1\mu_3}{\mu_1} + 1) \approx Z_2, \ \epsilon = \frac{a_0}{a}, \ \frac{1}{\mu_{12}} = \frac{1}{\mu_1} + \frac{1}{\mu_2}.$$

$\epsilon V(\mathbf{x},\mathbf{p},\epsilon)$ is Hill's potential of the Coulomb restricted three body problem. In order to describe the meaning of the symbols in the last example imagine three particles with charges e_1, e_2, e_3 and masses μ_1, μ_2, μ_3 so that $e_1e_2 < 0, e_1e_3 < 0$. I will consider the socalled lunar problem in which $|e_3|$ is small compared with $|e_1|, |e_2|$. This situation can be realized, e.g., if particle 1 is a negative ion, particle 2 is a positive ion and particle e_3 is an electron, so that $e_1 = Z_1e, \ e_2 = -Z_2e, \ e_3 = -e$, where e = charge of a positron and $Z_1, Z_2 \gg 1$. In any case I assume that e_3 does not influence the motion of particles one and two which are assumed to move on (semi)-classically on circular Kepler orbits.

In our **Coulomb restricted three body problem** one obtains to second order the following normal forms:

$$K_q = \omega(2J+1) - \gamma\epsilon^{3/2}\frac{1}{\omega}(2J+1)L_3(1+\frac{9}{8}\frac{\kappa^2}{\gamma^2\omega^2}\epsilon)+$$

$$\frac{(2J+1)}{8\omega^3}\epsilon^3[20Z_2J(J+1) - 6Z_2\mathbf{L}^2 - 15Z_2R_3^2 + (3Z_2-4\gamma^2)L_3^2],$$

$$K_c = 2\omega J - \gamma\epsilon^{3/2}\frac{1}{\omega}2JL_3(1+\frac{9}{8}\frac{\kappa^2}{\gamma^2\omega^2}\epsilon)+$$

$$\frac{2J}{8\omega^3}\epsilon^3[20Z_2J^2 - 6Z_2\mathbf{L}^2 - 15Z_2R_3^2 + (3Z_2-4\gamma^2)L_3^2].$$

In Kummer [1983] I have used the classical normal form to study the gravitational lunar problem ($\beta = 0$). In particular, I have described four families of quasiperiodic solutions with two frequencies and discussed their physical meaning.[16] This problem was also studied by van der Meer and Cushman [1986] by an alternative method of normalization starting from Moser's regularization (Moser [1970]) of the Kepler problem. One can use the quantum mechanical normal form to obtain the formula for the energy - eigenvalue splitting

$$E_{n,\ell,m} = E_n^0 - [Z_1Z_2\frac{\mu_3}{\mu_{12}}]^{1/2}(\frac{a_0}{a})^{3/2}[1+\frac{9}{8}n^2\frac{Z_2}{Z_1^3}(\frac{Z_1\mu_3}{\mu_1}+1)^2\frac{\mu_{12}}{\mu_3}\frac{a_0}{a}]\frac{e^2}{a_0}m$$

$$+n^2\frac{Z_2}{8Z_1^2}(\frac{a_0}{a})^3\frac{e^2}{a_0}[5(n^2-1) - 6\ell(\ell+1) - 15 < R_3^2 > +3m^2],$$

$E_n^0 = -\frac{Z_1^2e^2}{2a_0n^2}$ =Rydberg formula for hydrogen - like atoms. Here $< R_3^2 >$ signifies the expectation value of the the square of the third component of the Laplace - vector in the unperturbed energy eigen - state $|n,\ell,m>$ which can be expresssed in terms of the so called Glebsch Gordan coefficients of the rotation group. This result was derived from Formula (7.4.38), p. 344 in Biedenharn and Louck [1981] and after some manipulation can be brought into the following form

$$< R_3^2 >= \frac{1}{(2\ell-1)(2\ell+3)}[(n^2+m^2)((2\ell(\ell+1)-1) - 2n^2m^2 + 1 - 2\ell^2(\ell+1)^2]. \tag{123}$$

Second order Stark effect:
Quantummechanical and classical normal form of the Hamiltonian up to third order terms in the normalized variables are:

$$K_q = \omega(2J+1) + \frac{3}{2}\beta\frac{1}{\omega^2}(2J+1)R_1 - \frac{\beta^2}{16\omega^5}(2J+1)[51R_1^2 + 17(2J+1)^2 - 9L_1^2 + 19].$$

$$K_c = 2J\omega + \beta\frac{3}{\omega^2}JR_1 - \frac{\beta^2}{16\omega^5}2J[51R_1^2 + 17(2J)^2 - 9L_1^2].$$

An analysis of the classical normal form Hamiltonian shows that there are two families of quasi periodic solutions with two frequencies.

In the case of the (quantummechanical) Stark effect the states $|n,\ell,m>$ turn out not to be the correct eigenstates. Rather they are states $|j,\nu_1>|j,\nu_2>$ in which M_1, N_1, J are diagonal, i.e. states of the product representation $(j)\times$

[16]See also Figure 6 of the present paper.

$(j) = (2j) \oplus (2j-1) \oplus \ldots \oplus (0)$ of $SU(2)$. Using these the quantum normal form can be employed to reproduce the old result (see, e.g., Bethe and Salpeter [1957]) according to which the energy spectrum to second order is

$$E_{n,k,m} = -\frac{e^2}{2a_0 n^2} + \frac{3}{2} e a_0 k n \mathcal{E} - \frac{\mathcal{E}^2 a_0^3 n^4}{16}[17n^2 - 9m^2 - 3k^2 + 19],$$

where $k = \nu_1 - \nu_2$, $m = \nu_1 + \nu_2$, $n = 2j + 1$.

Quadratic Zeeman effect

$$H = H_0 + \alpha L_3 + \frac{\alpha^2}{2}(x_1^2 + x_2^2), \quad \alpha = B\frac{a_0}{e^2} b_0,$$

where $b_0 = \frac{eh}{2c\mu}$ = Bohr magneton, B= magnetic field.

The quantum normal form is:

$$K_q = \omega(2J+1) + \frac{\alpha}{\omega}(2J+1)L_3 + \frac{\alpha^2}{2\omega^3}(2J+1)[\frac{5}{2}[(2J+1)^2 - R_3^2] - \frac{1}{2}(L_3^2+1) - 2\mathbf{L}^2]$$

$$-\frac{\alpha^4}{\omega^7}(2J+1)\{\frac{21}{64} + \frac{239}{192}R_3^4 + \frac{383}{96}(2J+1)^2 - \frac{123}{32}(2J+1)^2 R_3^2 - \frac{145}{96}L_3^2(2J+1)^2 +$$

$$\frac{7}{32}L_3^2 R_3^2 + \frac{7}{192}L_3^4 + \frac{499}{192}(2J+1)^4 - \frac{35}{96}L_3^2 - \frac{259}{96}R_3^2$$

$$-11V_1 \circ [\frac{13}{12}(2J+1)^2 J - \frac{7}{12}R_3^2] + \frac{23}{12}L_3^2 V_1 - \frac{157}{6}V_1 + 9V_1^2\},$$

where $V_1 = (M_1 N_1 + M_2 N_2)$ and $\circ$ denotes symmetrization. The last Hamiltonian leads to the same approximation to the energy spectrum as the one found by Delande and Gay [1984]. These authors employ a special projection method adapted to certain perturbations of the hydrogen atom.

The corresponding classical normal form is manufactured from K_q by replacing $2J+1$ by $2J$ and only retaining for each power of α the terms of highest homogeneity degree. The following discussion ignores the term in the normal form of order α^4. Like in the Coulomb restricted three body problem we find four families of precessing Kepler orbits (see Figure 6). The most interesting families are the two families of inclined elliptical orbits which exist for $0 < |v| < v_0$.

$$\mathbf{L} = 2J[\pm e^{i\phi} sgn(-v)\sqrt{\frac{|v|}{v_0} - v^2}, v] = \text{ angular momentum}$$

$$\mathbf{R} = 2J[e^{i\phi}\sqrt{|v|v_0 - v^2}, \pm\sqrt{(v_0 - |v|)(\frac{1}{v_0} - |v|)}] = \text{Laplace vector.}$$

The value of v_0 is $\sqrt{\frac{3}{5}}$ for the Coulomb restricted three body problem and $\frac{1}{\sqrt{5}}$ for the Zeeman effect. This corresponds to a critical inclination of $39^0 14'$ for the three body problem and $63^0 26'$ for the Zeeman effect. [17]

Our procedure applied to q.m. is able to reproduce the result in the literature (see, e.g., Schiff and Snyder [1939]. For a different algebraic method reproducing this result see Becker and Bleuler [1976].)

[17]For an other approach to the classical Zeeman effect using second order normalization see Coffrey, Deprit, and Williams [1986] .

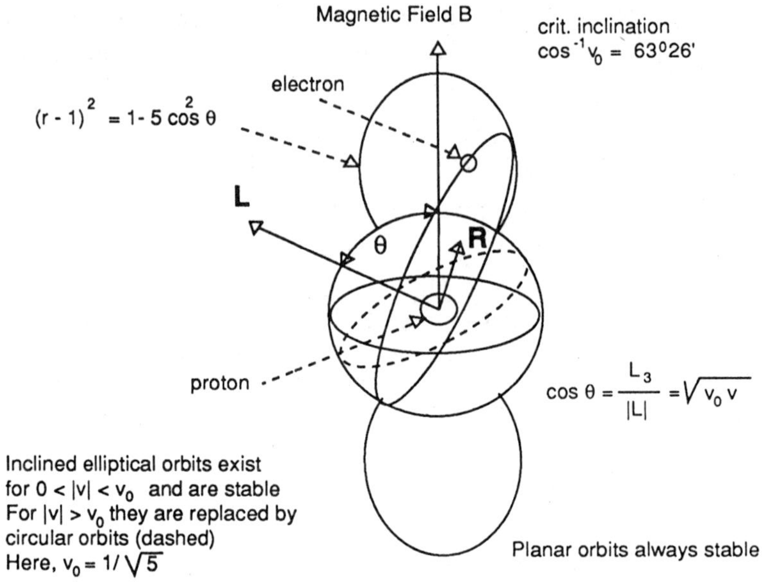

Figure 6 Quasiperiodic solutions in the classical Zeeman effect.

$$E_{n,\ell,m} = E_n^0 + Bb_0 m + B^2 \frac{a_0}{2e^2} b_0^2 n^2 \frac{[5n^2+1-3\ell(\ell+1)][\ell(\ell+1)-1+m^2]}{(2\ell-1)(2\ell+3)}.$$

Orthogonal superposition of Stark and Zeeman effect: In what follows we present the expressions for the classical and quantum normal forms. The classical expression differs slightly from the one found before us by van der Meer and Cushman [1987]. The source of this discrepancy at the time of this writing has not been resolved completely.

$$K_c = 2J\omega - \frac{2}{\omega} J\alpha L_3 + \frac{3}{\omega^2}\beta J R_1 - \frac{\alpha^2}{\omega^3} J L_3^2 + \frac{\alpha\beta}{\omega^4} J[7R_1L_3 - R_3L_1]$$

$$-\frac{\beta^2}{8\omega^5} J[51R_1^2 + 17(2J)^2 - 9L_1^2] + \mathcal{O}_3(\alpha,\beta).$$

$$K_q = \omega(2J+1) - (2J+1)(\frac{\alpha}{\omega}L_3 - \frac{3}{2}\frac{\beta}{\omega^2}R_1) - \frac{\alpha^2}{2\omega^3}(2J+1)L_3^2 +$$

$$\frac{\alpha\beta}{2\omega^4}(2J+1)[7R_1 \circ L_3 - R_3 \circ L_1]$$

$$-\frac{\beta^2}{16\omega^5}(2J+1)[51R_1^2 + 17(2J+1)^2 - 9L_1^2 + 19] + \mathcal{O}_3(\alpha,\beta).$$

$$E_{n,\ell,m}/\frac{e^2}{a_0} = -\frac{1}{2n^2} - \sqrt{\triangle}\frac{m}{2} + \frac{3}{16}\beta^2(9n^2\beta^2 - 14\alpha^2)\frac{m^2n^4}{\triangle} + \frac{15}{4}\alpha^2\beta^2\frac{n^4(n^2-1)}{\triangle} +$$

$$\frac{9}{16}\beta^2 n^4 \frac{2\alpha^2 + 9\beta^2 n^2}{\triangle} < R_3^2 > -\frac{9}{4}\alpha^2\beta^2 n^4 \frac{\ell(\ell+1)}{\triangle} - \frac{\beta^2}{16} n^4 [17n^2 + 19].$$

Here, $\triangle = 4\alpha^2 + 9\beta^2 n^2$ and $< R_3^2 >$ is given in Eq. (123).

6 Conclusion

I believe that I have amply demonstrated that there is much more overlap between classical and quantum mechanics than the traditional presentation in the textbooks reveals. Many important problems could be presented in a unified manner. At the core of such a presentation sits the notion of the normal form and of the Poisson algebra. In particular the approach to perturbations of the quantum mechanical Kepler problem that we have outlined above is worth pursuing in complete detail. Not only does it provide a clear cut algebraic procedure that avoids the evaluation of complicated integrals which is inherent in the traditional method, it may also lead to deeper physical insight into the quantum mechanical problem since it remains in close contact with the corresponding less abstract classical problem almost to the very end of the computational procedure. The approach also promises to shed new light on the algebraic relationship between classical and quantum mechanics.

References

Abraham, R., Marsden, J.E. [1978], *Foundation of Mechanics*, Benjamin Cummings, 276.

Ali, M.K. [1985], J. Math. Phys., **26**, 2565.

Bacry, H [1966], Il Nuovo Cimento, **41 A**, 222.

Baider, A., Churchill, R.C. [1988], Proc. R. Soc. Edinburgh A, **108**, 27.

Baider, A. [1989], J. Diff. Eq., **78**, 33.

Bargmann, V. [1936], Z. Physik, **98**, 169.

Bargmann, V. [1961], Comm. Pure Appl. Math., **14**, 187.

Bargmann, V. [1962], Rev. Mod. Phys., **34**, 829.

Barut, A. O. [1972], *Dynamical Groups and generalized Symmetries in Quantum Theory,* University of Canterbury Christchurch.

Barut, A. O., and Raczka, R. [1977], *Theory of Group Representations and Applications,* Polish Scientific Publishers, Warszaw.

Bates, L. [1988], Reports on Math. Phys., **26**, 413.

Bates, L. [1989], Reports on Math. Phys., **28**, 289.

Baumgarte, J. [1978], J. Reine und Angew. Math., **301**, 59.

Becker, H. G., and Bleuler, K. [1976], Z. Naturforschung, **31**, 517.

Bethe, H., and Salpeter, E.E. [1957], Handbuch der Physik, Vol., **35**, ed. S. Flügge, Springer-Verlag.

Biedenharn, L.C., and Louck, J. [1981], *Angular Momentum in Quantum Physics,* Encyclopedia of Mathematics and its Applications, **8**, Addison Wesley.

Churchill, R., Kummer, M., and Rod, D. [1983], J. Diff. Eqn., **49**, 359.

Churchill, R., and Lee, D. [1984], Lect. Notes in Pure and Appl. Math., **92**, 239.

Cushman, R., and Rod, D. [1982], Physica D, **6**, 105.

Coffey, S., Deprit, A., and Williams, C. A. [1986], Am. N.Y. Acad. Sci., **497**, 22.

Cushman, R. [1992], *Dynamics Reported,* **1** Jones, C.K.R.T. Kirchgraber, U., Walter, H. O., managing editors, Springer - Verlag, p. 54.

Delande, D., and Gay, J. C. [1984], J. Phys. B: At. Mo. Phys. **17**, L335.

Deprit, A. [1969], Celest.Mech., **1**, 12.

Deprit, A. [1984], *The Big Bang and George Lemaitre,* D. Reidel, 151.

Englefeld, M. J. [1972], *Group theory and the Coulomb problem,* John Wiley.

Fock, V. [1935], Z. Physik, **36**, 145.

Gyoergyi, G. [1968], Il Nuovo Cimento, **3A**, 717.

Jost, R. *unpublished manuscript.*

Kirillov, A. A. [1976], *Elements of the theory of representations*, Springer-Verlag.

Kummer, M. [1971], Il Nuovo Cimento, **1B**, 123.

Kummer, M. [1976], Commun. Math. Phys., **48**, 53.

Kummer, M. [1979], Amer. J. of Math., **101**, 6, 1333.

Kummer, M. [1982], Commun. Math. Phys., **84**, 133.

Kummer, M. [1983], Journal of Math. Anal. Appl., **93**, 142.

Kummer, M. [1984], *Proceedings of the 13th International Colloquium on Group Theoretical Methods in Physics,* World Scientific, p. 58.

Kummer, M. [1985], Archive for Rat. Mech. and Anal., **91**, 55.

Kummer, M. [1991], *Essays in Classical and Quantum Dynamics,* A.W. Saenz "Festschrift" volume, Gordon Breach, p.139.

Kummer, M., and Gompa, R. [1988], J. Math. Physics, **29**, 1405.

Kustaanheimo P., and Stiefel, E.L [1965], J. Reine Angew. Math., **218**, 204.

Ito H. [1989], Comment. Math. Hel., **64**, no. 3, 412.

Iwai, T., and Uwano, Y. [1986], J. Math., **27**, 1523.

Iwai, T., and Uwano, Y. [1988], J. Phys. A: Math. Gen., **21**, 4083.

Lemlih, A. B., and Ellison, J. A. [1985], Phys. Rev. Lett., **55**, 1950.

Marsden J. and Weinstein, A. [1974], Rep. Math. Phys., **5**, 121.

McIntosh, H. V. [1971], *Group Theory and its Applications,* **II**, (Loebl, E., ed.) Academic Press, 75.

Milnor, J. [1983], Amer. Math. Monthly, **90**, 353.

Mladenov, M. I. [1987], J. Phys. A Math. Gen., **20**, 17, 5865.

Moser, J. [1968], Memoirs of the Amer. Math. Soc., **81**, 1.

Moser, J. [1970], Comm. Pure Appl. Math. **23**, 609.

Moser, J. [1980], Progress in Mathematics, **8**, 233.

Pauli, W. [1929], Z. Physik, **36**, 336.

Schiff, L. I., and Snyder, H. [1939], Phys. Rev., **55**, 59.

Souriau, J. M. [1970], *Structure des Systèmes dynamiques,* Dunod, Paris.

Stiefel, E.L., and Scheifele G. [1971], *Linear and Regular Celestial Mechanics,* Springer.

Uwano, Y. [1989], Physica D, **35**, no. 1-2, 1.

van der Meer, J.C. [1985], Lect. Notes in Math., **1160**, Springer.

van der Meer, J. C., and Cushman, R. [1986], J. Appl. Math. and Phys. (ZAMP), **37**, 402.

van der Meer, J. C., and Cushman, R. [1987], *Differential geometric methods in theoretical physics,* (H. Doebner, ed.), World Scientific, p.403.

Vivarelli, M. D. [1985], Celestial Mechanics, **36**, 349.

Wood, R., and Ali, M. K. [1987], J. Phys. A: Math Gen., **20**, 351.

Wulfram, C. E. [1971], *Group Theory and its Applications,* **II**, (Loebl, E., ed.) Academic Press, 145.

Fields Institute Communications
Volume 8, 1996

On the Global Flow in the Coulombian Restricted Isosceles 3-body Problem

Ernesto A. Lacomba and J. Guadalupe Reyes
Universidad Autónoma Metropolitana
Departamento de Matemáticas
Av. Michoacán y la Purísima
Apdo. Postal 55-534
CP. 09340, México, D.F.

1 Introduction

During the last years there has been a renewed interest in studying the problems of Celestial Mechanics, whose equations of motion were discovered by Isaac Newton. Newton established the universal law of gravitation which describes the force of attraction between any two bodies.

The central problem in Celestial Mechanics is the n-body problem, that is, the motion of n particles with masses m_i $(i = 1, 2, ..., n)$ respectively, which move under the influence of their mutual gravitational forces. If r_{jk} is the distance between the k-th and the j-th particles then $\gamma m_j m_k / r_{jk}^2$ is the value of the attracting force on either of the two particles.

A repulsive n-body problem is obtained from the Newtonian attractive problem when the sign is changed in the potential of the equations of motion. That is, if n punctual bodies of masses $m_1, m_2, \ldots, m_n$ are moving in the Euclidean space $\mathbb{R}^N$, under these repulsive forces, the equations for the system are given by

$$\ddot{r}_i(t) = \frac{1}{m_i} \sum_{j \neq i} \frac{\gamma m_i m_j}{|r_i - r_j|^3}(r_i - r_j) \qquad i, j = 1, 2, \ldots, n$$

where $r_k(t)$ is the position at physical time t of the particle with mass m_k in the space $\mathbb{R}^N$ and $\gamma > 0$ is the (Coulombian) repulsion constant between the masses.

Using the corresponding Hamiltonian, it is easy to see that the energy is always positive in this system. Even more, there are no singularity (collision) solutions in this problem (see Peredo [1984]), all solutions are defined for any real time and all the velocities are bounded.

From the fact that energy is always positive, it follows that most of the particles must escape (Peredo [1984]), and the question is about the behavior of those which do not do that, having instead an asymptotic motion to an equilibrium position

1991 *Mathematics Subject Classification.* Primary: 34C35, 70F07; Secondary: 34A25, 58F05.

close to the center of mass. The conjecture is, that at most one particle can have such behavior.

Other studies have been made on particular cases of this problem.

Gal'perin [1981] and Moser [1977] have considered the case where the system has a central repulsion, that is, there exists a function $V : \mathbb{R} \to \mathbb{R}$ known as the effective potential such that the potential $U = U(r)$ of the system satisfies

a) $U(r) = V(|r|)$ (U is a central potential).

b) $V(x)$ is a monotone nonincreasing function for $x > 0$.

They have found that with some boundedness and integrability conditions at infinity for V, one has the following asymptotic behavior at ∞:

$$r_i(t) = c_i + v_{i,\infty} \cdot t + o(1), \qquad \text{as} \qquad t \to \infty$$

where c_i are vectors determined uniquely by the initial conditions and $v_{i,\infty}$ the velocity at ∞ of the particle with mass m_i.

Peredo [1984] proved that in the collinear n-body problem with repulsive forces the two exterior bodies escape to ∞ and $-\infty$ respectively.

We consider here the problem obtained when we have a 3-particle system moving under the influence of repulsive coulombian forces and having always an isosceles triangle configuration.

In this paper we describe the global flow for the restricted Coulombian isosceles 3-body problem by using geometrical methods for the study of escapes, as we did in Lacomba and Reyes [1991] and [1995], but studying the solutions of another type of singular complex second order differential equations. In those papers we analyzed the solutions in a singular point of differential equations of the type $z\omega'' = F(\omega', \omega, z)$, obtained after we applied a blow up transformation. We use here the same blow up, but we analyze the solutions in a singular point of a differential equation of the type $z^2\omega'' = F(z\omega', \omega, z)$, obtaining in a simpler way the analytical description of the escapes in the direction of the central configurations. However, for searching escapes in the other directions we use the same type of differential equations as in Lacomba and Reyes [1991], but we change the value of the parameter in the blow up proposed there. The complete global flow is given at any positive time for solutions whose initial conditions project into a closed half-strip contained in Hill's region. If such projection falls elsewhere in the Hill's region, some restrictions apply, and the global description has to be done in a less precise way. We use standard analysis in the interior of the Hill's region for obtaining a global description of the solutions, generalizing Peredo's [1984]. And we use new techniques of blow up for studying the solutions touching the boundary curve of such configuration region. We do that by studying the analytical solutions of a differential equation of the type $\omega'' = F(\omega', \omega, z)$.

For more details of some particular repulsive Coulombian problems, the reader is referred to Peredo [1984] and Lacomba and Peredo [1989] where a motivation of the problem is given. Ramírez [1993] has studied the 3-body collinear problem by using similar methods to the ones used here.

We use Jacobi-like coordinates with center of mass at the origin and taking two particles with equal mass m and the third with mass μ. Then, according to Figure

1, the motion is described by the equations

$$\begin{cases} \ddot{x} &= \dfrac{\gamma m}{4x^2} + \dfrac{\gamma\mu x}{(x^2+y^2)^{3/2}} \\ \\ \ddot{y} &= \dfrac{\gamma(\mu+2m)y}{(x^2+y^2)^{3/2}} \end{cases} \tag{1.1}$$

where $\gamma > 0$ is a constant of repulsion between the particles.

They can be written as a Hamiltonian system, with Hamiltonian function

$$\mathcal{H} = \frac{1}{2}\left[2m\dot{x}^2 + \frac{2\mu}{2m+\mu}\dot{y}^2\right] + \frac{\gamma m^2}{2x} + \frac{2\gamma m\mu}{(x^2+y^2)^{1/2}}.$$

In particular, $\mathcal{H}$ is a first integral, because if we use the equation (1.1) it is easy to verify that $\frac{d\mathcal{H}}{dt} = 0$, that is, $\mathcal{H} = h$ is constant along the solutions. Then it is sufficient to study the solutions in a fixed level of energy h (always positive).

By Sard's Theorem, we can suppose the set $\{(x, y, \dot{x}, \dot{y}) | \mathcal{H} = h\}$ is a 3-dimensional manifold in the phase space.

If we set the third mass equal to zero $\mu = 0$, we obtain the restricted Coulombian problem.

$$\begin{cases} \ddot{x} &= \dfrac{\gamma m}{4x^2} \\ \ddot{y} &= \dfrac{2\gamma m y}{(x^2+y^2)^{3/2}}, \end{cases} \tag{1.2}$$

which is not a Hamiltonian system any more. However it has the energy relation $\frac{\dot{x}^2}{2} + \frac{\gamma m}{4x} = h$ for the movement of the nonzero mass particle m. This first integral is obtained from $\mathcal{H}$ letting $\mu = 0$ and dividing by $2m$.

For this reason we use here slightly different techniques than in the study of Hamiltonian systems.

To let $\mu = 0$ corresponds in practice to the nonrestricted case when the mass of the third particle is small enough.

For this problem there are three simple solutions, the so-called homothetic or central configurations

$y \equiv 0$ (Euler's solution or collinear)

$y \equiv \pm\sqrt{3}x$ (Lagrange's solutions or equilateral).

In this problem the particles with mass m always escape and for the motion of the zero mass particle Peredo [1984] obtained the following result.

Theorem 1.1 *(Zero mass particle). For the initial conditions $x_0 = x(0)$, $y_0 = y(0)$, $\dot{x}_0 = \dot{x}(0)$, $\dot{y}_0 = \dot{y}(0)$ we have that*

i) *If $y_0 > 0$ and $\dot{y}_0 \geq 0$ then $\lim_{t\to\infty} y(t) = \infty$*

ii) *If $y_0 > 0$, then exists a velocity $u < 0$ such that if $\dot{y}_0 = u$, then $\lim_{t\to\infty} y(t) = L$ for some value $L \geq 0$.*

Proof: By Taylor's Theorem exists $\xi \in (0, t)$ such that $y(t) = y(0) + \dot{y}(0)t + \frac{\ddot{y}(\xi)}{2}t^2$.

For the case $y(0) > 0$ and $\dot{y}(0) \geq 0$ we claim that $y(t) > 0$ for all $t > 0$. Assume there exists a first positive time $\bar{t}$ such that $y(\bar{t}) = 0$.

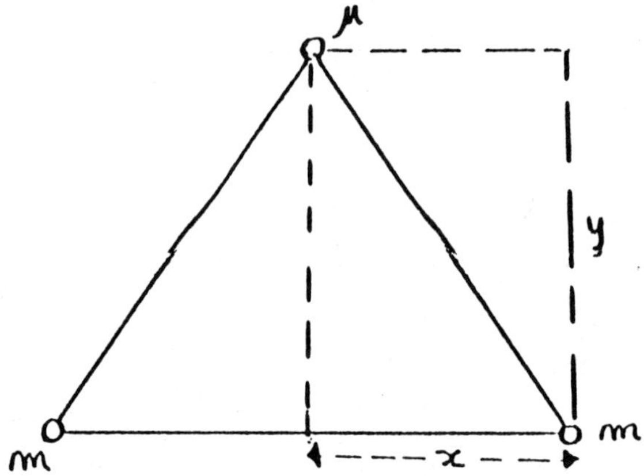

Figure 1 The isosceles 3-body problem in Jacobi coordinates.

By Taylor's Theorem $0 = y(\bar{t}) = y(0) + \dot{y}(0)\bar{t} + \frac{\ddot{y}(\xi)\bar{t}^2}{2}$, for some $\xi \in (0, \bar{t})$. This implies that $\ddot{y}(\xi) < 0$. This is a contradiction to $y(\xi) > 0$.

This short analysis indicates that if $y(0) > 0$ and $\dot{y}(0) \geq 0$ then $y(t) > 0$ for $t > 0$, in particular for $\xi \in (0, t)$.

With this $\ddot{y}(\xi) > 0$ because the sign in the right hand side of the second equation in (1.2) is the same as the sign of y. From this $y(t) \geq y(0) + \dot{y}(0)t$, which implies necessarily that $\lim_{t\to\infty} y(t) = \infty$ if $y(0) > 0$ and $\dot{y}(0) \geq 0$.

By symmetry, if $y(0) < 0$ and $\dot{y}(0) \leq 0$ then $\lim_{t\to\infty} y(t) = -\infty$.

The above shows that Euler's solution $y \equiv 0$ is unstable.

Using standard analysis as above (see Peredo [1984]) it can be shown in a similar way that if $y(0) > 0$ and $\dot{y}(0) < 0$ then

a) If $|\dot{y}(0)|$ is small enough, then there exists $t^* > 0$ such that $y(t^*) > 0$, $\dot{y}(t^*) = 0$, and hence $\lim_{t\to\infty} y(t) = \infty$.

b) If $|\dot{y}(0)|$ is big enough, then there exists $t^* > 0$ such that $y(t^*) < 0$, $\dot{y}(t^*) \leq 0$, that is, $\lim_{t\to\infty} y(t) = -\infty$.

Therefore, there exist two negative velocities u_1 and u_2 such that $u_1 < u_2$ and if $\dot{y}(0) = u_1$ then $\lim_{t\to\infty} y(t) = -\infty$, but if $\dot{y}(0) = u_2$ then $\lim_{t\to\infty} y(t) = +\infty$.

We consider the sets

$$\begin{aligned} \Lambda_+ &= \{\dot{y}(0) \in [u_1, u_2] |\, \lim_{t\to\infty} y(t) = \infty\} \\ \Lambda_- &= \{\dot{y}(0) \in [u_1, u_2] |\, \lim_{t\to\infty} y(t) = -\infty\}. \end{aligned}$$

We claim that Λ_+ and Λ_- are relative open sets in $[u_1, u_2]$, and that there exists $u \in [u_1, u_2]$ such that $y(t)$ does not tend to $+\infty$ nor to $-\infty$ as $t \to \infty$.

Let us denote by $y(t; y_0, v)$ the solution $y(t)$ of (1.2) for the initial conditions $x_0 = x(0)$, $\dot{x}(0) = 0$, $y_0 = y(0)$ and $\dot{y}(0) = v$.

We note that if $v_0 \in \Lambda_+$ then there exists $t_1 > 0$ such that $y(t_1; y_0, v_0) > y_0$ and $\dot{y}(t_1; y_0, v_0) > 0$.

Continuity respect to initial conditions in bounded intervals implies uniform continuity of any solution $y(t; y_1, v)$ of (1.2) on the compact set

$K = [0, t_1] \times [0, y(t_1; y_0, v_0)] \times [u_1, u_2]$. From this, we can verify that there exists $\delta > 0$ such that if $v' \in [u_1, u_2]$ and $|v' - v_0| < \delta$ then $v' \in \Lambda_+$. Therefore Λ_+ is relative open in $[u_1, u_2]$. In a similar way Λ_- is also relative open in $[u_1, u_2]$.

Since Λ_+ and Λ_- are non-empty subsets in $[u_1, u_2]$, and $\Lambda_+ \cap \Lambda_- = \emptyset$, then by the connectedness of $[u_1, u_2]$ it follows that there exists $u \in [u_1, u_2]$ such that $u \notin \Lambda_+$ and $u \notin \Lambda_-$. We now claim that $\lim_{t\to\infty} y(t; y_0, u) = L$ for some $L \geq 0$.

We have that $\dot{y}(0; y_0, u) < 0$ and $\dot{y}(t; y_0, u) > 0$ for some $t > 0$ is not possible because $u \notin \Lambda_+$. Therefore $\dot{y}(t; y_0, u) \leq 0$ for all $t \geq 0$, which implies that $y(t; y_0, u)$ is decreasing on $[0, \infty)$. From this it follows that $\lim_{t\to\infty} y(t; y_0, y) = L$ where $L < \infty$.

If $L < 0$ then $\lim_{t\to\infty} y(t; y_0, u) = L = -\infty$. This is not possible because $u \notin \Lambda_-$, hence $L \geq 0$.

□

Proposition 1.1 *If* $0 < y(0) < \frac{x(0)}{\sqrt{2}}$ *and* $\dot{x}_0 \geq 0$, *then there exists a unique velocity* $u < 0$ *such that*

a) *If* $\dot{y}_0 > u$ *then* $\lim_{t\to\infty} y(t) = \infty$

b) *If* $\dot{y}_0 < u$ *then* $\lim_{t\to\infty} y(t) = -\infty$

c) *If* $\dot{y}_0 = u$ *then* $\lim_{t\to\infty} y(t) = L$ *for some value* $L \geq 0$.

Proof: Assume that $y_1(t)$ and $y_2(t)$ are two solutions with $y_1(0) = y_2(0) = y_0$ and $\dot{y_1}(0) \leq \dot{y_2}(0) < 0$ and with finite limits $y_1(t) \to L_1$ and $y_2(t) \to L_2$ as $t \to \infty$. For fixed t, the acceleration $H(y) = \frac{2\gamma m y}{(x(t)^2+y^2)^{\frac{3}{2}}}$ is increasing if $y < \frac{x(t)}{\sqrt{2}}$. Hence $(y_2 - y_1)^{\cdot\cdot}(t) \geq 0$ if $t \geq 0$.

Since $(y_2 - y_1)(0) = 0$, $(y_2 - y_1)^{\cdot}(0) \geq 0$ and $(y_2 - y_1)(t) \to L_2 - L_1$ as $t \to \infty$, this is impossible, unless $y_1(t) \equiv y_2(t)$.

□

Proposition 3.1 below will show that in general $L > 0$.

2 Asymptotic and Parallel escapes.

2.1 Transformation of Coordinates In this section we describe the possibilities of escape of the particles in the Coulombian repulsive restricted isosceles 3-body problem, as the physical time goes to infinity.

To begin with we observe that the first equation of system (1.2) is uncoupled from the second one, corresponding to the movement of the nonzero mass particles as a collinear repulsive problem without considering the third one. By using classical methods of integration we can obtain the following result.

Proposition 2.1 *(Nonzero mass particles). If we introduce the uniformizing time variable* τ *such that* $\frac{dt}{d\tau} = \frac{x}{\sqrt{\frac{h}{2}}}$, *then the solutions of the first equation in (1.2) are parametrized by the formulae*

$$x(\tau) = \frac{\gamma m}{8h}(1 + \cosh 2\tau)$$

$$t(\tau) = \frac{\gamma m}{8\sqrt{2h^3}}(2\tau + \sinh 2\tau).$$

Proof: We consider the energy relation for the first equation of system (1.2):

$$\frac{\dot{x}^2}{2} + \frac{\gamma m}{4x} = h. \tag{E}$$

then from $\frac{x}{\sqrt{\frac{h}{2}}} = \frac{dt}{d\tau}$, by the chain's rule $\dot{x} = \frac{\sqrt{\frac{h}{2}}}{x}\frac{dx}{d\tau} = \frac{\sqrt{\frac{h}{2}}}{x}x'$, and substituting in the energy equation we have the relation

$$(x')^2 + \frac{\gamma m x}{h} = 4x^2.$$

Differentiating this last equation with respect to τ we obtain $x'' - 4x + \frac{\gamma m}{2h} = 0$ with general solution

$$x(\tau) = a\cosh(2\tau) + b\sinh(2\tau) + \frac{\gamma m}{8h},$$

where a and b are arbitrary constants but because of the equation $(x')^2 - 4x^2 + \frac{\gamma m}{h}x = 0$ they have to satisfy the relation $a^2 - b^2 = \frac{\gamma^2 m^2}{64h}$. If $b = 0$ then $a = \frac{\gamma m}{8h}$, and therefore the general solution is given by $x(\tau) = \frac{\gamma m}{8h}(\cosh(2\tau) + 1)$

□

Corollary 2.1 *(Peredo [1984]). For any given initial conditions on the nonzero mass particles, we have*

$$\lim_{t\to\pm\infty} x(t) = \infty.$$

Morever, the parametric relations in Proposition 2.1 imply that $x \sim \sqrt{2h}|t|$ *if* $t \to \pm\infty$ *(See Lacomba and Peredo [1989]), which means that at infinity the physical variable* t *behaves asymptotically as the configuration variable* x.

If we fix a positive energy level $h > 0$, we obtain the Hill's region $M_h = \{(x,y)|x \geq \frac{\gamma m}{4h}\}$ as the set of possible configurations of the system (1.2); $\Gamma = \{(x,y)|x = \frac{\gamma m}{4h}\}$ is the boundary curve of the Hill's region.

It is clear that $(x,y) \in \Gamma$ if and only if $\dot{x} = 0$, from the energy relation (E).

Corollary 2.2 *If* $(x_0 = x(0), y_0 = y(0))$ *and* $(\dot{x}_0 = \dot{x}(0), \dot{y}_0 = \dot{y}(0))$ *are arbitrary initial conditions of problem (1.2), there exists a unique time* t^* *such that* $\dot{x}(t^*) = 0$. *Moreover, if* $t > t^*$ *then* $\dot{x}(t) > 0$.

Proof: From the parametric relations in Proposition 2.1 we have $\frac{dt}{d\tau} = \frac{x}{\sqrt{\frac{h}{2}}} > 0$, so, the function $t : \mathbb{R} \to \mathbb{R}$ is bijective. Therefore the inverse function $\tau = \tau(t)$ is defined in all $\mathbb{R}$ and it is of class C^∞, by the inverse function theorem.

On the other hand $x = x(\tau)$ is an even function, and therefore the composition $x = x(\tau(t)) = x(t)$ is an even function.

Moreover by Corollary 2.1, $x(t) \to +\infty$ as $t \to \pm\infty$. By the mean value theorem there exists $t^* \in \mathbb{R}$ such that $\dot{x}(t^*) = 0$. The uniqueness follows from the fact that $\ddot{x}(t) > 0$ for all $t \in \mathbb{R}$.

□

We conclude from this corollary that every solution of (1.2) with arbitrary initial conditions intersects Γ and does it only once. On the other hand, if along a solution of (1.2) the nonzero mass particles reach zero velocity, then it will be necessarily intersecting Γ in the Hill's region.

Since any solution $(x, y) = (x(t), y(t))$ intersects Γ, it can be seen as the graph of two functions $y = y(x)$ in the Hill's region M_h, defined in $(\frac{\gamma m}{4h}, \infty)$. Hence, we can replace the physical time t by the variable x for studying the behaviour of the zero mass particle as $t \to \pm\infty$, using this way the configuration variable x as the new independent variable. This is valid for $t > t^*$ or for $t < t^*$ in the terminology of Corollary 2.2.

From the chain's rule, for $x > \frac{\gamma m}{4h}$ we have that

$$\dot{y} = \frac{dy}{dx}\dot{x} \qquad \text{and} \qquad \ddot{y} = \frac{d^2y}{dx^2}\dot{x}^2 + \frac{dy}{dx}\ddot{x},$$

which when we substitute in the second equation (1.2) and consider the energy relation (E), we obtain the differential equation

$$2\big(h - \frac{\gamma m}{4x}\big)\frac{d^2y}{dx^2} + \frac{dy}{dx}\big(\frac{\gamma m}{4x^2}\big) = \frac{2\gamma m y}{(x^2+y^2)^{3/2}}, \tag{2.1}$$

whose solutions are the orbits without parametrization corresponding to the solutions of (1.2).

Proposition 2.2 *(Angle of a solution at ∞) Peredo [1984]. Let $(x = x(t),\, y = y(t))$ be a solution of (1.2) for some initial condition, and define by $\theta(t) = \arctan \frac{y(t)}{x(t)}$ the angle of configuration at time t, then*

i) *$\theta_\infty = \lim_{t\to\infty} \theta(t)$ exists and satisfies $\frac{-\pi}{2} < \theta_\infty < \frac{\pi}{2}$.*

ii) *Moreover $\theta_\infty = \arctan\big(\frac{v_\infty}{\sqrt{2h}}\big)$, where $v_\infty = \lim_{t\to\infty}\dot{y}(t)$ is the velocity at ∞ of the zero mass particle, which is well defined.*

Proof: From the energy relation $\dot{x}^2 + \frac{\gamma m}{2x} = 2h$, we obtain that $\lim_{t\to\infty}\dot{x}(t) = \sqrt{2h}$.

Using an elaborated analysis (Peredo [1984]), we can verify that for any initial condition for the zero mass particle, the velocity at infinity exists and it is bounded: $\lim_{t\to\infty}\dot{y}(t) = v_\infty$.

By using the L'Hôpital rule we have that

$$\theta_\infty = \lim_{t\to\infty} \arctan\left(\frac{y(t)}{x(t)}\right) = \lim_{t\to\infty} \arctan\left(\frac{\dot{y}(t)}{\dot{x}(t)}\right) = \arctan\left(\frac{v_\infty}{\sqrt{2h}}\right)$$

□

So that every solution of (1.2) in M_h has a limiting slope. We shall seek asymptotic motions in the Hill's region for the solutions $y = y(x)$ of (2.1). By eliminating the physical time t we have that by Proposition 2.2, $\lim_{x\to\infty}\arctan\big(\frac{y(x)}{x}\big)$ exists and is bounded. From here, the search for solutions of (2.1) asymptotic to some straight line with slope α when $x \to \infty$, should be detected by the projective transformation of coordinates $M_h \to H \times \mathbb{R}P^1$ defined by $(x, y) \to (x, u) = (x, y/x)$, where $H = \{x | x \geq \frac{\gamma m}{4h}\}$ and $\mathbb{R}P^1$ is the real projective line of dimension 1.

By a straightforward calculation, we obtain from the relation $y = ux$:

$$\frac{dy}{dx} = u + x\frac{du}{dx} \qquad \text{and} \qquad \frac{d^2y}{dx^2} = x\frac{d^2u}{dx^2} + 2\frac{du}{dx},$$

and the equation (2.1) is transformed by the projective transformation into the differential equation

$$2\left(h-\frac{\gamma m}{4x}\right)x\frac{d^2u}{dx^2}+\left(4h-\frac{3\gamma m}{4x}\right)\frac{du}{dx}+\frac{\gamma mu}{4x^2}=\frac{2\gamma mu}{x^2(1+u^2)^{3/2}} \tag{2.2}$$

defined in $H\times\mathbb{R}P^1$.

Now we take the inversion $v=\frac{1}{x}$ and using once more the chain's rule, we obtain in the new variables (v,u) the second order differential equation

$$v^2\frac{d^2v}{du^2}=\left[\frac{4\gamma mu}{(4h-\gamma mv)(1+u^2)^{3/2}}+\frac{\gamma m(v\frac{du}{dv}-u)}{2(4h-\gamma mv)}\right]v \tag{2.3}$$

defined in $V\times\mathbb{R}P^1$, where $V=\{v\in\mathbb{R}|0\le v\le\frac{4h}{\gamma m}\}\subset\mathbb{R}$.

Hence the composite projective inversion transformation $M_h\to V\times\mathbb{R}P^1$ defined by $(x,y)\to\left(\frac{1}{x},y/x\right)=(v,u)$ shall detect every asymptotic solution of (2.1) to the straight line $y=\alpha x+\beta$ as the solution of (2.3) which crosses the line $\{0\}\times\mathbb{R}P^1$ at the point $(0,\alpha)$ with a slope β. This is because the transformation carries the straight lines $y=\alpha x+\beta$ in M_h, into straight lines $u=\beta v+\alpha$ in $V\times\mathbb{R}P^1$ exchanging slopes by origin ordinates and vice versa. In Lacomba and Reyes [1991] we have studied the relation between the attractive and repulsive problems by considering the solution of (2.3) on the set $V=\{v\in\mathbb{R}||v|\le\frac{4h}{\gamma m}\}$. We are interested in looking for solutions of (2.3) which cross the line $\{0\}\times\mathbb{R}P^1$ with the initial conditions $v=0$, $u(0)=\alpha$ and $\frac{du}{dv}(0)=\beta$. Since $v=0$ is a singularity of the differential equation (2.3), to get analytic solutions, we shall use analytic continuation by complexifying the variables (v,u) to $(z,\omega)\in\Omega\times\mathbb{C}P^1$, where $\Omega\subset\mathbb{C}$ is a region of the complex plane containing V and $\mathbb{C}P^1$ is the one dimensional complex projective line.

We then get the second order differential equation on $\Omega\times\mathbb{C}P^1$

$$z^2\omega''=\left[\frac{4\gamma m\omega}{(4h-\gamma mz)(1+\omega^2)^{3/2}}+\frac{\gamma m(z\omega'-\omega)}{2(4h-\gamma mz)}\right]z, \tag{2.4}$$

where $'$ denotes the derivative respect to z.

We observe than the equation (2.4) is of the form $z^2\omega''=F(z\omega',\omega,z)$, where F is an holomorphic function defined in an open set of $\mathbb{C}\times\mathbb{C}P^1\times\Omega$ and $z=0$ is a singularity of the differential equation.

A necessary condition for (2.4) to have analytic solutions of the form $\omega(z)=\alpha+\beta z+\dots$ for the initial conditions $z=0$, $\omega(0)=\alpha$, $\omega'(0)=\beta$, is that $F(0,\alpha,0)=0$ and it satisfies the identity $(F_{100}+F_{010})\beta=-F_{001}$, where $F_{100}=\frac{\partial F}{\partial(z\omega')}(0,\alpha,0)$, $F_{010}=\frac{\partial F}{\partial\omega}(0,\alpha,0)$ and $F_{001}=\frac{\partial F}{\partial z}(0,\alpha,0)$. (See Forsyth [1900]).

But $F(0,\alpha,0)=0$ is an identity, and a calculation shows that $F_{100}=F_{010}\equiv 0$, so β will be arbitrary if and only if $F_{001}=0$ is satisfied. In fact $F_{001}=\frac{\gamma m\alpha}{h(1+\alpha^2)^{3/2}}-\frac{\gamma m\alpha}{8h}$ vanishes only for the values $\alpha=0,\pm\sqrt{3}$.

On the other hand, a sufficient condition for the existence of analytical solutions for those initial conditions is that the critical quadratic associated to the differential equation (2.4) does not have positive integers as roots. The critical quadratic in this case is $\lambda^2-\lambda(F_{100}-3)+(2-2F_{100}-F_{010})=0$, see Forsyth [1900].

Since $F_{100}\equiv 0$ and $F_{010}\equiv 0$ the critical quadratic becomes $\lambda^2+3\lambda+2=0$, whose roots are $\lambda=-1,-2$ which are nonpositive integers.

We summarize the above in the following:

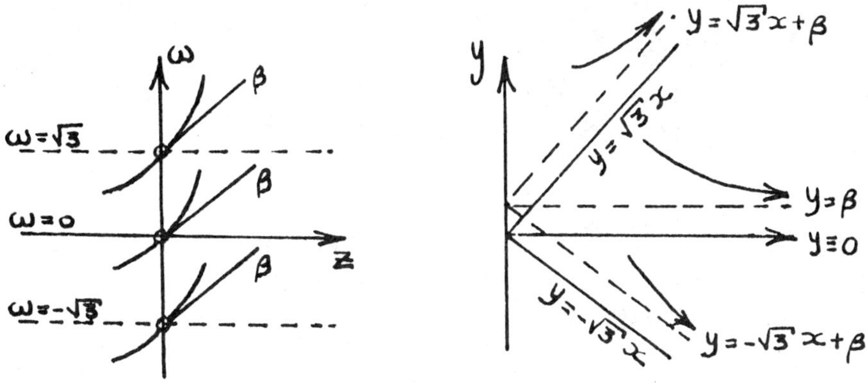

Figure 2 Parallel solutions asymptotic to the central configurations.

Theorem 2.1 *The differential equation (2.4) defined in $\Omega \times \mathbb{C}P^1$ has analytic solutions $\omega = \omega(z)$ which cross the line $\{0\} \times \mathbb{C}P^1$, if and only if the initial conditions are $z = 0$, $\omega \in \{0, \pm\sqrt{3}\}$ and $\omega' = \beta$ is arbitrary. Morever, these analytic solutions are unique for the given initial conditions.*

When we restrict to $V \times \mathbb{R}P^1$ we obtain the following.

Proposition 2.3 *The differential equation (2.3) has unique analytic solutions which cross the line $\{0\} \times \mathbb{R}P^1$ if and only if $\alpha \in \mathbb{R}P^1$ satisfies that $v(0) = \alpha \in \{0, \pm\sqrt{3}\}$ and $v'(0) = \beta$ is arbitrary.*

We observe that for $\alpha \in \{0, \pm\sqrt{3}\}$ the straight line $u = \alpha$ is surrounded by a family of solutions of (2.3) of the form $u = \alpha + \beta v + \cdots$ in every direction β, crossing $\{0\} \times \mathbb{R}P^1$ at the point $(0, \alpha)$. The solutions of (1.2) in the Hill's region corresponding to such solutions have slope α at infinity, and they are asymptotically parallel to the directions of the central configurations. The solutions $y = y(x)$ of (2.1) converge asymptotically to some straight line of the form $y = \alpha x + \beta$ because the solutions corresponding to (2.3) in $V \times \mathbb{R}P^1$ cross $\{0\} \times \mathbb{R}P^1$, tangentially to the straight line $u = \beta v + \alpha$ at the point $(0, \alpha)$.

If we summarize the last analysis we obtain:

Theorem 2.2 *For every direction of central configuration α in the space of configuration there exists in the 4-dimensional phase space a 2-dimensional analytic submanifold formed by velocity vectors of solution curves of (1.2) such that when projected in the Hill's region they become parallel to the straight line $y = \alpha x$ as $t \to \infty$.*

Figure 2 illustrates Theorem 2.2.

2.2 Ramified solutions at infinity. We answer now the question of whether asymptotic escape in the other directions can also be detected by a transformation of the type considered above for the central configurations. This is possible by introducing a ramified transformation to be defined below. The ramification implies that we have to use a singular differential equation which is slightly different from the one above. Incidentally, if in the equation (2.4) we cancel a factor z, its type would be the same as that of the equation (2.5) below. This was analyzed by the authors Lacomba and Reyes [1991], but it presents some difficulties because the critical quadratic has 0 as a double root.

We will use the new transformation of coordinates

$$\tilde{\Omega} \times \mathbb{C}P^1 \longrightarrow \Omega \times \mathbb{C}P^1$$

defined by $(\tau, \omega) \to (\tau^\rho, \omega) = (z, \omega)$, where $\tilde{\Omega}$ is some ramified covering of Ω under the function $\tau \to \tau^\rho$, where $\rho > 1$ is not a positive integer.

From the change of coordinates $z = \tau^\rho$ we obtain the equalities

$$\omega' = \frac{\tau^{1-\rho}}{\rho}\frac{d\omega}{d\tau} \quad , \quad \omega'' = \frac{\tau^{2-2\rho}}{\rho^2} \cdot \frac{d^2\omega}{d\tau^2} + \frac{1-\rho}{\rho^2}\tau^{1-2\rho}\frac{d\omega}{d\tau},$$

which carry the differential equation (2.4) into the new equation in the variable (τ, ω):

$$\tau\ddot{\omega} = \frac{4\gamma m \rho^2 \tau^{\rho-1}\omega}{(4h - \gamma m \tau^\rho)(1+\omega^2)^{3/2}} + \frac{\gamma m \tau^{\rho-1}(\frac{\tau}{\rho}\dot{\omega} - \omega)}{2(4h - \gamma m \tau^\rho)} + (\rho - 1)\dot{\omega} \tag{2.5}$$

defined in $\tilde{\Omega} \times \mathbb{C}P^1$. Here $(\cdot)$ denotes the derivative with respect to τ.

The differential equation (2.5) can be treated in a similar way as the equation (2.4) for finding solutions $\omega = \omega(\tau)$ to the initial conditions $\tau = 0$, $\omega = \alpha$, $\dot{\omega} = \beta$. (See Forsyth [1900]).

Theorem 2.3 *If $\rho > 1$ is not a positive integer, then the differential equation (2.5) has solutions $\omega = \omega(\tau)$ analytical in the variables τ, $\tau^{\rho-1}$ crossing the line $\{0\} \times \mathbb{R}P^1$ if and only if the initial conditions are $\tau = 0$, $\omega(0) = \alpha$ arbitrary and $\dot{\omega} = \beta = 0$.*

Proof: Let $G(\dot{\omega}, \omega, \tau)$ be the analytic function on the right hand side of equation (2.5).

A necessary condition for having analytical solutions is that the equation $G(\beta, \alpha, 0) = 0$ holds. (See Forsyth [1900]). A simple substitution shows that such condition is equivalent to $(\rho - 1)\beta = 0$, or $\beta = 0$ and $\omega = \alpha$ is arbitrary.

By calculating the roots of the critical quadratic associated to (2.5), we obtain $\lambda(\lambda - g_1) = 0$, where $g_1 = \frac{\partial G}{\partial \dot{\omega}}(\beta, \alpha, 0)$ and such quadratic has roots $\lambda = 0$ and $\lambda = \frac{\partial G}{\partial \dot{\omega}}(\beta, \alpha, 0) = \rho - 1$.

It can be shown in this case that, for the initial conditions $\tau = 0$, $\omega = \alpha$, $\dot{\omega} = \beta = 0$ there are an infinity of solutions all of them analytic in the variables τ, $\tau^{\rho-1}$. (See Forsyth [1900]). In fact, the general form of the solution for such initial conditions is

$$\omega(\tau) = \alpha + a_{\rho-1}\tau^{\rho-1} + \sum_{m,n \geq 2}^{\infty} a_{mn}\tau^{m(\rho-1)}\tau^n, \tag{2.6}$$

an analytic function in the above arguments, and where $a_{\rho-1}$ is an arbitrary constant, (See Forsyth [1900]).

□

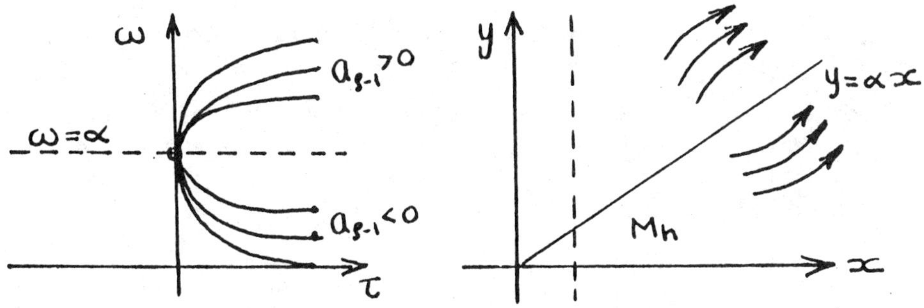

Figure 3 Parallel escapes in the direction of α.

In order to describe the behaviour of escapes in the solution of equation (1.2) corresponding to the solutions of (2.5), we first observe the geometry of the transformation $M_h \to \tilde{\Omega} \times \mathbb{R}P^1$ defined by $(x, y) \to \left(\frac{1}{x^{1/\rho}}, \frac{y}{x}\right) = (\tau, \omega)$.

It is easy to see that the straight line $y = \alpha x + \beta$ is carried into the curve $\omega = \alpha + \beta\tau^\rho$ which has contact order ρ in the point $(0, \alpha)$ with the line $\omega = \alpha$; and one curve of the form $\omega = \alpha + \beta\tau^k$ is carried by the inverse into a curve of the form $y = \alpha x + \beta x^{1-k/\rho}$.

When a solution (2.6) is transformed to the variables (x, y) in the Hill's region, it takes the form

$$y = \alpha x + a_{\rho-1}x^{1/\rho} + \sum_{m,n\geq 2}^{\infty} a_{mn}x^{m(1/\rho-1)}x^{1-n/\rho}. \tag{2.7}$$

Lemma 2.1 *Let us fix a positive number* $1 < \rho < 2$ *and let it be a solution* $\omega = \alpha + a_{\rho-1}\tau^{\rho-1} + \ldots$ *as above. We have that*

1. *If* $a_{\rho-1} \neq 0$, *then the solution curve corresponding for the differential equation (1.2) in the Hill's region has a limiting slope* α, *but it is not asymptotic to any straight line* $y = \alpha x + \beta$ *as* $t \to \infty$.
2. *If* $a_{\rho-1} = 0$, *then the solution curve corresponding for the differential equation (1.2) in the Hill's region is asymptotic to the straight line* $y = \alpha x$ *as* $t \to \infty$.

Proof: Notice that all the terms in the summation of (2.7) go to zero as $x \to \infty$.

If $a_{\rho-1} \neq 0$ then $y = \alpha x + a_{\rho-1}x^{1/\rho} + \sum_{m,n\geq 2}^{\infty} a_{mn}x^{m(1/\rho-1)}x^{1-n/\rho}$ which has slope α at infinity but is not asymptotic to some straight line of the form $y = \alpha x + \beta$.

If $a_{\rho-1} = 0$ then $y = \alpha x + \sum_{m,n\geq 2}^{\infty} a_{mn}x^{m(1/\rho-1)}x^{1-n/\rho}$ which has an asymptotic straight line $y = \alpha x$ at infinity.

□

Figure 3 illustrates the Theorem 2.3 and Lemma 2.1.

Lacomba [1987] described the asymptotic behaviour of the orbits escaping for the n-body problem when the energy is positive, showing with precision how a neighborhood of infinity is foliated by invariant submanifolds formed by orbits escaping at infinity in every direction of the configuration escape. This applies equally well to repulsive problems. More precisely:

Theorem 2.4 *(Lacomba [1987]). For a given slope α, there exists in the phase space a 2-dimensional submanifold formed by velocity vectors solutions of (1.2), which when they are projected in the Hill's region, become parallel or asymptotic to the straight line $y = \alpha x$ as $t \to \infty$.*

In the ramified description (2.7), the parameter $a_{\rho-1}$ can take all the real values. This generates exactly the required 2-dimensional submanifold in the phase space formed by solutions escaping in the given direction. The difference with the case of the central configurations is that the solutions are defined everywhere except at $z = 0$, $\omega = \alpha$, which corresponds to the point at infinity in the direction of α in the original configuration space M_h. The reason is that the solution is analytic not in τ, but in the two variables τ and $\tau^{\rho-1}$. This is why $\dot{\omega}(0)$ is not defined in (2.6) for $a_{p-1} \neq 0$.

3 Motions in the Hill's Region

3.1 Global behaviour of solutions We begin by clarifying that the limiting value L of the vertical coordinate for the asymptotic motions in Theorem 1.1 and Proposition 1.1 is always positive.

Proposition 3.1 *(Lacomba and Reyes [1994]). If $(x_0, y_0, \dot{x}_0, \dot{y}_0)$ is any condition for the differential equation (1.2) such that $\lim_{t\to\infty} y(t) = 0$, then the solution for such initial configuration is the Euler solution $y \equiv 0$. For any other solution with initial condition $y_0 > 0$ the asymptotic value of the coordinate y is either $\pm\infty$ or a non-zero real number.*

Proof: If $y(t)$ has a finite positive limit L as t goes to ∞, the corresponding solution $(x(t), y(t))$ of (1.2) in the phase space is contained in the analytic submanifold formed by the solutions escaping in the direction of the Euler configuration. When we replace t by x, the corresponding solution to the differential equation (2.1) has the form

$$y(x) = L + \frac{a_2}{x} + \cdots .$$

The corresponding solution to equation (2.3) has the form

$$v(u) = Lu + a_2 u^2 + \cdots ,$$

which is analytic in $u = \frac{1}{x}$.

By the uniqueness of this solution according to Proposition 2.3, we have that $v(u)$ is the Euler solution in these coordinates if and only if $L = 0$ and $v \equiv 0$.

□

We observe that, in general, any vertical velocity u_a which carries the zero mass particle to an asymptotic equilibrium does depend on the initial velocity $\dot{x}_0 = a$ of the particles with nonzero mass. We denote the corresponding asymptotic value of the vertical coordinate by β_a. For example, if $\dot{x}_0 = \pm 1$, then $u_{-1} \neq u_1$ and $\beta_1 \neq \beta_{-1}$ in general.

Recall that

$$M_h = \left\{(x, y) | x \geq \frac{\gamma m}{4h}, y \in \mathbb{R}\right\}$$

is the Hill's region of the system, for fixed positive energy $h > 0$. We are interested in describing globally the flow at some points in the open upper half of such region $M_h^+ = \{(x, y,)|x > \frac{\gamma m}{4h}, y \geq 0\}$, which does not contain to the boundary curve $\Gamma = \{(x, y)|x = \frac{\gamma m}{4h}, y \in \mathbb{R}\}$.

The region M_h^+ is divided into the following three subsets, which we shall call *zones*:

$$A = \left\{(x, y) \in M_h^+|x > \frac{\gamma m}{4h}, 0 \leq y \leq \frac{\gamma m}{4\sqrt{2}h}\right\}$$

$$B = \left\{(x, y) \in M_h^+|x > \frac{\gamma m}{4h}, \frac{\gamma m}{4\sqrt{2}h} < y < \frac{x}{\sqrt{2}}\right\}$$

$$C = \left\{(x, y) \in M_h^+|x > \frac{\gamma m}{4h}, \frac{x}{\sqrt{2}} \leq y\right\}.$$

If $P(x_0, y_0) \in M_h^+$ then $|\dot{x}_0| \neq 0$, because the boundary curve Γ of M_h is where the velocities of the nonzero mass particles vanish.

Let us consider the tangent plane to M_h at the point P, denoted by $T_P M_h \cong \mathbb{R}^2$. For a fixed nonzero velocity $\dot{x}_0 = a$, we consider 4 types of tangent vectors in $T_P M_h$ as initial conditions at (x_0, y_0) for equations (1.2) pointing out in four different regions in the affine space $T_P M_h = P + \mathbb{R}^2$.

- Region I: $v = (a, \dot{y}_0) \in T_P M_h$ with $a > 0$, and $\dot{y}_0 > 0$.
- Region II: $v = (a, \dot{y}_0) \in T_P M_h$ with $a < 0$, and $\dot{y}_0 > 0$.
- Region III: $v = (a, \dot{y}_0) \in T_P M_h$ with $a < 0$, and $\dot{y}_0 < 0$.
- Region IV: $v = (a, \dot{y}_0) \in T_P M_h$ with $a > 0$ and $\dot{y}_0 < 0$.

In the next subsection we describe the flow at $P(x_0, y_0) \in A$ by analyzing the four possible types of initial conditions at such point. For the behaviour of the flow in *zones* B and C, see the remark below Theorem 3.1.

Similar results hold for the case when the initial condition is $y_0 < 0$, because of the symmetry $y \to -y$ in equations (1.2).

3.2 Motions in the interior of Hill's region We describe here the behaviour of solutions through a point $P(x_0, y_0)$ in the half-strip A. That is, in this subsection we suppose $x_0 > \frac{\gamma m}{4h}$ and $\frac{\gamma m}{4\sqrt{2}h} \geq y_0 > 0$.

Notice that Proposition 1.1 holds in *zone* A. That is, given $P(x_0, y_0) \in A$ and $\dot{x}_0 = a \neq 0$, there exists a unique initial velocity u_a for the zero mass particle, which satisfies a), b) and c) of that Proposition.

We recall from Corollary 2.2 that if $P(x_0, y_0)$, $Q(\dot{x}_0, \dot{y}_0)$ is an arbitrary initial condition of the problem (1.2), then there exists a unique time t^* such that $\dot{x}(t^*) = 0$. Moreover if $t > t^*$ then $\dot{x}(t) > 0$. Hence all solutions of (1.2) with any initial conditions, intersect the curve Γ only once before going to infinity (escaping).

We consider a fixed velocity $a \neq 0$ for the point $P(x_0, y_0) \in A$. The following result follows directly from Proposition 1.1 a).

Corollary 3.1 *Let $P(x_0, y_0) \in M_h$ be as before, and $v = (a, \dot{y}_0) \in T_P M_h$ with $\dot{x}_0 > 0$ and $\dot{y}_0 > 0$, then $\lim_{t\to\infty} x(t) = \lim_{t\to\infty} y(t) = \infty$.*

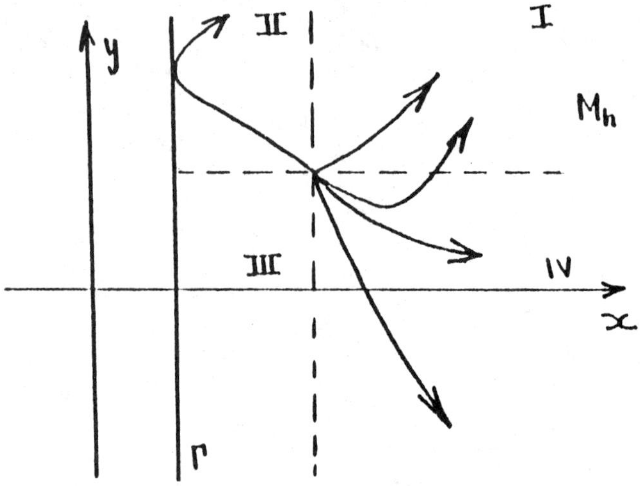

Figure 4 Motions in Regions I,II, IV for a point in A.

This means that, for initial conditions pointing into region I, the particles escape. At infinity the corresponding curve in the Hill's region is of linear order as graph of a function $y = y(x)$, as we saw in Section 1. Now we describe the region IV, as a particular case of Proposition 1.1.

Corollary 3.2 *If $(a, \dot{y}_0) \in T_pM_h$ is such that $a > 0$, $\dot{y}_0 < 0$, there exists a unique negative velocity $u_a = u(\dot{x}_0)$ such that*

a) *If $0 > \dot{y}_0 > u_a$ then $\lim_{t\to\infty} x(t) = \infty$ and $\lim_{t\to\infty} y(t) = \infty$.*

b) *If $\dot{y}_0 < u_a$ then $\lim_{t\to\infty} x(t) = \infty$ and $\lim_{t\to\infty} y(t) = -\infty$.*

c) *If $\dot{y}_0 = u_a$ then $\lim_{t\to\infty} x(t) = \infty$ and $\lim_{t\to\infty} y(t) = \beta_a > 0$ for some β_a.*

Corollaries 3.1 and 3.2 contain the results obtained by Peredo [1984] and Lacomba and Peredo [1989], but in the more general case, that is $\dot{x}_0 > 0$. In both cases, for no $t > 0$, the aforementioned solutions intersect Γ.

We define the zero velocity curve $\Gamma_0 = \{(x_0, y_0, \dot{x}_0, \dot{y}_0) | \dot{x}_0 = \dot{y}_0 = 0\}$ contained in the phase space, and whose projection into Hill's region is Γ.

Lemma 3.1 *If $v = (\dot{x}_0, \dot{y}_0) \in T_PM_h$ is such that $\dot{y}_0 \geq 0$ and $\dot{x}_0 < 0$, then the solution of (1.2) passing through P with tangent vector v intersects Γ in a positive time t^*, but the lifting to tangent vectors of such a solution does not intersect the zero velocity curve Γ_0, and $\lim_{t\to\infty} x(t) = \infty$, $\lim_{t\to\infty} y(t) = \infty$.*

Proof: From Corollary 2.2 the projection of the solution of (1.2) through P intersects Γ for a unique time $t^* > 0$ and with velocity $(0, \dot{y}(t^*))$. From the fact that $\dot{y}(0) \geq 0$ it follows that $\dot{y}(t^*) > 0$, and then the solution does not intersect at all the zero velocity curve Γ_0. From Theorem 1.1., it follows that $\lim_{t\to\infty} x(t) = \infty$ and $\lim_{t\to\infty} y(t) = \infty$.

□

The three preceding results describe the projection of solutions through a point $(x_0, y_0) \in A$ with tangent vectors pointing into the regions I, II and IV. Figure 4 depicts the situation for positive time $t > 0$.

We describe now the motions for the orbits through $P \in A$ with tangent vector pointing into the region III. This case gives a greater variety of situations.

From Theorem 1.1 for the fixed negative velocity $\dot{x}_0 < 0$, there exists a unique negative velocity u_a such that, if $0 > \dot{y}_0 > u_a$ ($\dot{x}_0 = a$) then $\lim_{t\to\infty} y(t) = \infty$. Hence $\dot{y}(t^*) > 0$ for big enough positive time t^*. By the Intermediate Value Theorem, there exists a unique time $t_{\dot{y}_0} = t(\dot{y}_0)$ such that $\dot{y}(t_{\dot{y}_0}) = 0$ where $\dot{y}_0 < 0$ is any initial velocity in $(u_a, 0]$. That is, $t_{\dot{y}_0}$ is the time when the zero mass particle backs up after starting with negative velocity $\dot{y}_0$.

This defines a function $t\colon (u_a, 0] \to [0, \infty)$ by $t(\dot{y}_0) = t_{\dot{y}_0}$. Obviously this function is continuous by the continuity with respect to initial conditions of problem (1.2), it is injective, and satisfies that $t(0) = 0$, $\lim_{\dot{y}_0 \to u_a} t(\dot{y}_0) = \infty$. This proves the following result.

Lemma 3.2 *There exists a continuous bijective function $t\colon (u_a, 0] \to [0, \infty)$ such that to any velocity $\dot{y}_0 \in (u_a, 0]$, it associates the time when the zero mass particle backs up to ∞, after having initiated its motion with velocity $\dot{y}_0$.*

Now, let us keep fixed the negative velocity $\dot{x}_0 < 0$, and let t^* be the unique time such that $\dot{x}(t^*) = 0$. By Lemma 3.2 there exists a unique negative velocity $z_a \in (u_a, 0]$ such that $t(z_a) = t^*$. Let us form the initial conditions $P(x_0, y_0)$ and $Q(\dot{x}_0, z_a)$ for the differential equation (1.2), with the above values. Then $\dot{y}(t^*) = \dot{x}(t^*) = 0$, and we obtain the following result.

Proposition 3.2 *For the point $P(x_0, y_0)$ with fixed horizontal velocity $\dot{x}_0 < 0$ there exists only one vertical velocity $z_a < 0$, such that the tangent vectors of the solution of (1.2) through P with initial velocity $Q(\dot{x}_0, z_a)$ intersect the zero velocity curve Γ_0 in a positive time, and the projection intersects Γ transversally.*

Let u_a be the velocity in the Theorem 1.1. We summarize all the results in Proposition 3.1, Lemma 3.2 and Proposition 3.2, in the following theorem for motions into the region III, determined by the critical velocities z_a and u_a.

Theorem 3.1 *For fixed $\dot{x}_0 < 0$, let z_a, u_a be the critical velocities as before for the problem (1.2) at the point $(x_0, y_0) \in A$, then the following alternatives hold for positive times*

1. *If $0 > \dot{y}_0 > z_a$, then $\lim_{t\to\infty} y(t) = \infty$, where the velocity of the zero mass particle vanishes before the velocity of the nonzero mass particle does. From this, the solution of (1.2) under these conditions does not intersect Γ_0.*

2. *If $\dot{y}_0 = z_a$, then $\lim_{t\to\infty} y(t) = \infty$, where the velocities of all the particles vanish simultaneously for some positive time. From this, the corresponding solution intersects Γ_0, the projection intersects Γ and backs up onto itself, due to the reversibility of the system.*

3. *If $z_a > \dot{y}_0 > u_a$, then $\lim_{t\to\infty} y(t) = \infty$, and in this case the velocities of the nonzero mass particles vanish before the zero mass particle velocity does. The corresponding solution does not intersect Γ_0.*

4. *If $\dot{y}_0 = u_a$ then $\lim_{t\to\infty} y(t) = \beta_a > 0$. The velocity of the zero mass particle never vanishes, and the corresponding solution does not intersect Γ_0.*

5. *If $\dot{y}_0 < u_a$ then $\lim_{t\to\infty} y(t) = -\infty$. The velocity of the third particle never vanishes, and such solution does not intersect Γ_0. (See Figure 5).*

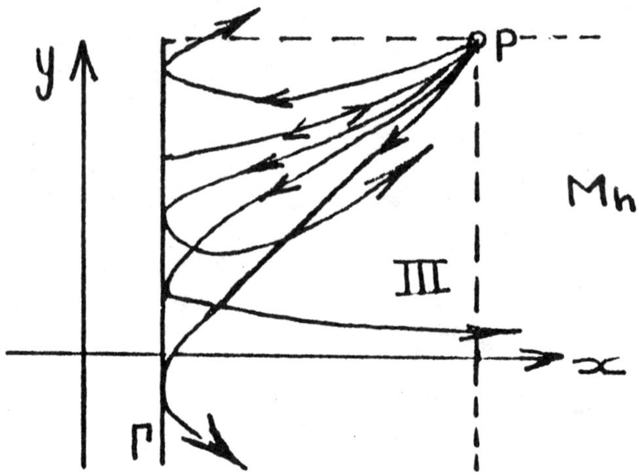

Figure 5 Motions on Region III for a point in A.

Remark. We describe now the behaviour of the flow in the other *zones*.

Consider the *zone*

$$B = \left\{ (x, y) \in M_h^+ | x > \frac{\gamma m}{4h}, \frac{\gamma m}{4\sqrt{2}h} < y < \frac{x}{\sqrt{2}} \right\}.$$

The Corollaries 3.1, 3.2, Lemma 3.1 and Proposition 3.2 are still valid. But in this case we can not assert the uniqueness of the critical velocity u_a in the Theorem 3.1.

On the other hand, consider the *zone*

$$C = \left\{ (x, y) \in M_h^+ | x > \frac{\gamma m}{4h}, \frac{x}{\sqrt{2}} \leq y \right\}.$$

The Corollary 3.1, Lemma 3.1 and Proposition 3.2 are still valid. But as in the above case we can not assert either the uniqueness for the respective critical velocities u_a in Corollary 3.2 and Theorem 3.1.

4 Blow up at the Boundary curve of the Hill's region

In this section we give a description of the way the solutions intersect the boundary Γ of the Hill's region in problem (1.2). We construct a new blow up transformation which helps us in understanding such behaviour. This transformation extends the flow, unfolding the solutions intersecting Γ, and constructing a fiber bundle over this boundary curve to describe the way solutions intersect it.

The results in Section 2 show that any solution of (1.2) can be seen (from a given time on or before that given time) as the graph of a function $y = y(x)$, except when the curve Γ is reached.

We extend the description of solutions to the boundary of Hill's region by doing the following covering change of coordinates $\mathbb{R}^2 \to M_h$ defined by $(z, \omega) \to (x, y)$ where $z^2 = x - \frac{\gamma m}{4h}$, $\omega = y$.

We observe that under this transformation, Γ is carried into the axis $\{0\} \times \mathbb{R} \subset \mathbb{R}^2$.

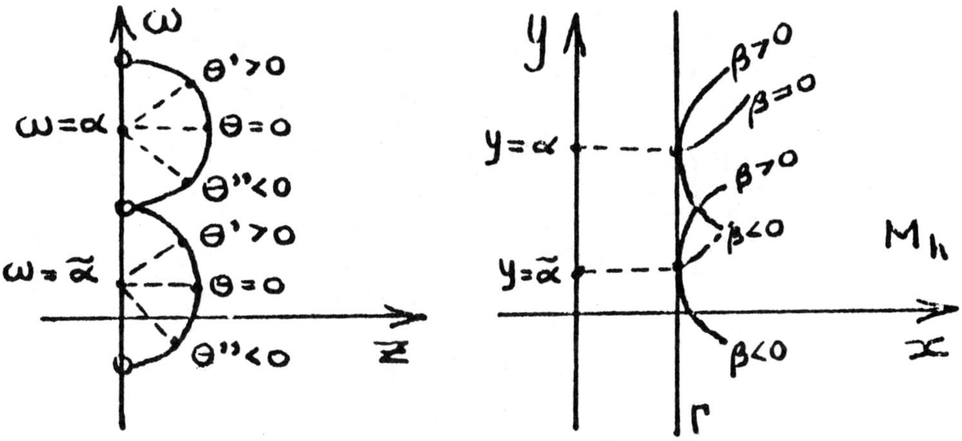

Figure 6 The correspondence between the bundle and the projected solutions along Γ.

In this variable (z, ω), the differential equation (2.1) becomes

$$\omega'' = \frac{\gamma m(4hz^2 + \gamma m)}{h\left[\left(z^2 + \frac{\gamma m}{4h}\right)^2 + \omega^2\right]^{3/2}} + \left[\frac{4hz}{4hz^2 + \gamma m}\right]\omega' \tag{4.1}$$

defined in the whole plane $\mathbb{R}^2$. This equation (4.1) has the type $\omega'' = F(\omega', \omega, z)$ with analytical right hand side. The classical Fuch's result (Forsyth [1900]) states that all the solutions of this equation are analytical.

So, for such given initial conditions $z = 0$, $\omega = \alpha$, $\omega' = \beta$, the general solution of (4.1) is given by the series

$$\omega = \alpha + \beta z + \sum_{n=2}^{\infty} a_n z^n, \tag{4.2}$$

which converges in a neighborhood of $z = 0$, where a_n does depend on α and β.

Lemma 4.1 *The space of local solutions of the differential equation (4.1) which cross the line* $\{0\} \times \mathbb{R}$ *is diffeomorphic to the product* $\mathbb{R} \times \left(-\frac{\pi}{2}, \frac{\pi}{2}\right)$.

Proof: To cross the line $\{0\} \times \mathbb{R}$ means $z = 0$, $\omega = \alpha$ and $\omega' = \beta$ arbitrary. From Mather's Determination Theorem (Mather [1968]) via local diffeomorphisms around the domains of $\omega = \omega(z)$ and $z = 0$, any solution (4.2) of (4.1) through $(0, \alpha)$ can be carried into the form $\omega = \alpha + \beta z$. Then we give the diffeomorphism $\phi : \alpha + \beta z \to (\alpha, \theta)$, where $\beta = \tan\theta$.

□

Since the initial derivative directions $\lambda = \theta$ and $-\lambda = \theta + \pi$ correspond to the same solution for the differential equation (4.1) through $(0, \alpha)$, we observe that if we consider the identification $\lambda \sim -\lambda$, the quotient space of the classes of directions of $[0, 2\pi]$ is $\mathbb{R}P^1$, where we exclude the direction of the axis $\{0\} \times \mathbb{R}$. That is, we can construct a fiber bundle $\mathcal{F}$ on the line $\{0\} \times \mathbb{R}$ where every fiber is the interval $I = (-\pi/2, \pi/2) \cong \mathbb{R}P^1 - \{\text{point}\}$.

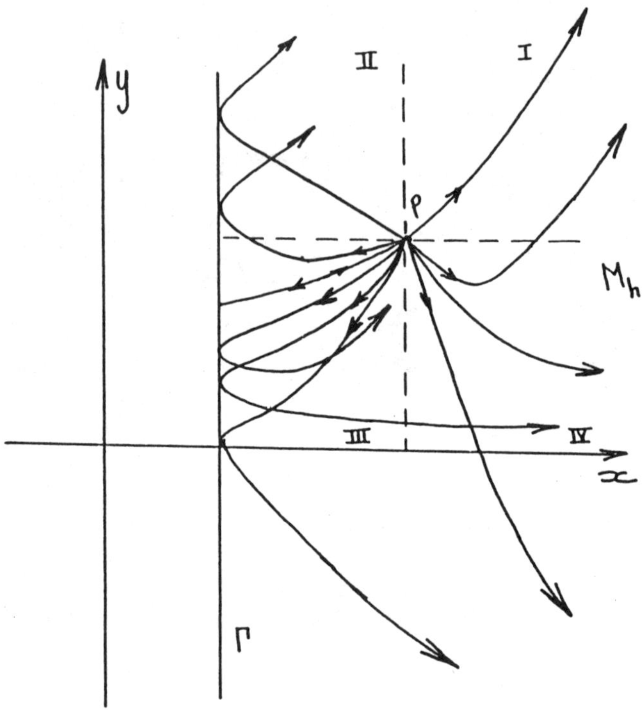

Figure 7 The global Flow in zone A.

In other words, $\mathcal{F} = \mathbb{R} \times I$ is a trivial bundle (Cartesian product) on the curve $B = \{0\} \times \mathbb{R}$, where the fibers (factors) are diffeomorphic to the interval $I = (-\pi/2, \pi/2)$.

Definition. The trivial fiber bundle $(\mathcal{F}, B)$ is called the unfolded blow up of the curve Γ with respect to differential equation (1.2).

When we write the general solution (4.2) of (4.1) in the original variables (x, y) in the Hill's region, we obtain that the general solution along Γ is of the form

$$y = \alpha + \beta \left(x - \frac{\gamma m}{4h}\right)^{1/2} + \sum_{n=2}^{\infty} a_n \left(x - \frac{\gamma m}{4h}\right)^{n/2}. \tag{4.3}$$

The radical can take the positive or the negative sign, corresponding to the description of the solution just before or just after touching the curve Γ.

Here we distinguish two cases:

If $\beta = 0$, then $y = \alpha + \sum_{n=2}^{\infty} a_n \left(x - \frac{\gamma m}{4h}\right)^{n/2} = \alpha + a_2 \left(x - \frac{\gamma m}{4h}\right) + a_3 \left(x - \frac{\gamma m}{4h}\right)^{3/2} + \ldots$, a curve with contact order bigger than 1 at $(0, \alpha)$ with the line $y = \alpha$.

If $\beta \neq 0$ then $y = \alpha + \beta \left(x - \frac{\gamma m}{4h}\right)^{1/2} + \ldots$, a curve which has a ramification of order $\frac{1}{2}$ at the point $(0, \alpha)$.

Figure 6 describes the aforementioned cases. The last construction carries via the diffeomorphisms, the trivial bundle over $\{0\} \times \mathbb{R}$ into a family of curves along the line Γ, with contact order equal to zero (or tangential to Γ) if $\beta \neq 0$, and with contact order greater or equal than one (or transversal to Γ) if $\beta = 0$. That is, if we consider the zero section of the bundle $\mathcal{F}_0 = \{(\alpha, 0) | \alpha \in \{0\} \times \mathbb{R}$ and

$0 \in \mathbb{R}P^1\}$, it is carried under the aforementioned construction into the projected curves that intersect the zero velocity curve Γ_0, and have contact order ≥ 1 with Γ. If β is considered as a parameter, the above considerations can be summarized in the following result.

Theorem 4.1 *Let $P(x_0, y_0)$ be a point in Γ. Then there exists a one parameter family of tangent vectors of solutions through P in the phase space of the differential equation (1.2), such that all except one of them become tangent to the boundary curve Γ. Such a solution intersects the zero velocity set Γ_0 in the phase space.*

Remark. Llibre [1983] used blow up techniques to replace the zero velocity curve of a Hamiltonian system with two degrees of freedom by a 2-dimensional manifold: the so-called zero velocity manifold. In such a manifold, every point of the zero velocity curve blows up into a circle, that is, the zero velocity manifold is locally the product of an interval (of the curve) and the circle: $I \times S^1$.

The Hamiltonian system extends in a smooth way to this manifold, obtaining a flow in the augmented space, giving an easy description of the flow near the zero velocity curve.

The reason for not using the techniques developed in Llibre [1983], is that the system studied here is not Hamiltonian and the curve Γ does not correspond to points in M_h, where $\dot{x} = \dot{y} = 0$ always holds, but it is only the boundary of the Hill's region.

We have described completely the flow of (1.2) along the boundary curve Γ of the Hill's region. Hence, we have a complete description of the global flow in the restricted Coulombian isosceles 3-body problem in the closed half-strip cl(A). (See Figure 7).

References

Forsyth A.R. [1900], *Theory of differential equations,* **III**, Dover.

Galperin, Yu. [1981], *The asymptotic behavior of a system of particles with repulsion,* Russ. Math. Surveys, **36**, 5, 147–148.

Lacomba, E.,A. [1987], *Infinite manifolds of positive energy in Celestial Mechanics,* Contemp. Math., **58**, Part III, 193–201.

Lacomba, E. A. and Peredo, F. [1989], *Escape-equilibrium solutions in the repulsive Coulombian isosceles 3-body problem,* Geometrie Symplectique et Mécanique, Lec. Notes in Math. **1416**, Springer-Verlag, 178–191.

Lacomba, E. A. and Reyes, J.G. [1991], *Unbounded parallel and asymptotic solutions in the isosceles 3-body problem with inverse-square forces,* Centre de Recerca Matematica, Bellaterra. (Barcelona), Spain, Report No. 130.

Lacomba, E. A. and Reyes, J.G. [1995], *Parallel escapes in the restricted repulsive Coulombian isosceles 3-body problem,* in New Trends in Hamiltonian Systems and Celestial Mechanics, World Scientific Publications, Singapore, 263-273, to appear.

Llibre, J. [1983], *A note on the energy levels of Hamiltonian systems of two degrees of freedom,* Actas del VI CEDYA, Jaca, Spain, 403–408.

Mather, J. [1968], *Stability of C^∞ mappings III. Finitely determined map germs,* Inst. Hautes Etudes Sci., Publ. Math., **35**, 127–156.

Moser, J. [1977], *The scattering problem for some particle systems on the line,* Springer Verlag, Lecture Notes in Math. **597**, 441–463.

Peredo, F. [1984], *The restricted isosceles problem with repulsive forces,* (Spanish), Ph. D. dissertation at the CINVESTAV, México D.F.

Ramírez, R. [1993], *The collinear repulsive restricted 3-body problem,* (Spanish), Master's thesis, UNAM, México, D.F.

Fields Institute Communications
Volume 8, 1996

Reductive Group Actions on Kähler Manifolds

Eugene Lerman
Mathematics Department
University of Illinois at Urbana-Champaign
1409 W. Green Street
Urbana, Illinois , USA 61801-2975

Reyer Sjamaar
Department of Mathematics
Cornell University
Ithaca, New York USA 14853-7901

Much attention has been given to the problem of "dividing out" a space by a group of transformations in such diverse areas of mathematics as classical mechanics and algebraic geometry. In classical mechanics one typically deals with a phase space M, a submanifold $L \subset M$ defined as a level set of a number of integrals of motion, plus a group G of canonical transformations leaving the motion on L invariant. The "reduced" phase space is then the space L/G. Modern treatments of this construction have been given by Meyer [1973] and Marsden and Weinstein [1974].

Algebraic geometers like to study complex-projective varieties M acted upon by groups of projective transformations that are usually not compact. Often, though, their groups are reductive, i.e., they are complexifications $G^{\mathbf{C}}$ of compact real Lie groups G. A good notion of a quotient in this category has been defined by Mumford, Fogarty and Kirwan [1994]. It consists in choosing a subset M^{ss} of "semistable" points of M, where the action is "good", and then forming a quotient $M^{\mathrm{ss}}/\!\!/G^{\mathbf{C}}$. This quotient is smaller than the naive, set-theoretical quotient $M^{\mathrm{ss}}/G^{\mathbf{C}}$, which is not a Hausdorff space owing to the fact that the orbits of the non-compact group $G^{\mathbf{C}}$ are not closed. The space $M^{\mathrm{ss}}/\!\!/G^{\mathbf{C}}$ is basically the set of *closed* orbits in M^{ss}.

Complex-projective manifolds are not classical phase spaces, but they have very special symplectic forms, obtained by restricting the Fubini-Study form on the ambient projective space. This raises the question whether there is any relation between the symplectic and the algebro-geometric quotients of such spaces. Guillemin, Sternberg, Mumford and others have made the important observation

1991 *Mathematics Subject Classification.* Primary: 58F06; Secondary 14L30, 19L10.
The authors were partially supported by an NSF grant.
An expanded version will appear elsewhere.

that for projective manifolds the notions of a quotient in symplectic and algebraic geometry are essentially the same. Kempf and Ness [1979] have proved an analogous result for affine varieties. See Bott [1988] for an account of this story. On the quantum level this correspondence leads to integral formulae for multiplicities of representations; see Guillemin and Sternberg [1982].

We have sought to generalize these multiplicity formulae to the case where the reduced phase space L/G is not a manifold, but a singular space. In the process we had to investigate more closely the correspondence between quotients in mechanics and in algebraic geometry, and the properties of actions of reductive complex Lie groups on Kähler manifolds. In particular, we had to address the issue of the existence of slices for such actions. The problem of linearizing the action near a fixed point and the construction of slices has been solved by Luna [1973] for affine varieties and by Snow [1982] for Stein spaces. The difficulty of the problem lies in the fact that an action of a reductive group $G^{\mathbf{C}}$ is never proper, unless it is locally free. One therefore faces the challenge of controlling the behaviour of the action "at infinity in the group". It seems unlikely to us that this can be done in general without imposing some restrictions on the space M.

In this note we present an alternative proof of a theorem of Koras [1986] to the effect that one can linearize a $G^{\mathbf{C}}$-action at a fixed point on a Kähler manifold M. The advantage of our proof is that it can be generalized to show the existence of slices at certain points of M, namely the points m such that the compact orbit Gm through m is *isotropic*. Slices and multiplicity formulae for singular reduced phase spaces will be the subject of a forthcoming paper. (See also Heinzner and Loose [1994].) As a foretaste we will at the end of this paper spell out an example given by Luna, showing that slices do not exist at arbitrary points of M. The result to be discussed here is the following:

Theorem 1 *Let M be a Kähler manifold and let $G^{\mathbf{C}}$ act holomorphically on M. Assume the action of the compact real form G is Hamiltonian. Let m be any fixed point of the $G^{\mathbf{C}}$-action. Then the action of $G^{\mathbf{C}}$ can be linearized in a neighbourhood of m, i.e., there exist a $G^{\mathbf{C}}$-invariant open neighbourhood U of m in M, a $G^{\mathbf{C}}$-invariant open neighbourhood U' of the origin 0 in the tangent space T_mM and a biholomorphic $G^{\mathbf{C}}$-equivariant map $U \to U'$.*

We find ourselves unable to understand part of Koras' proof of this theorem. In particular, we fail to see a justification for his application of the curve selection lemma. Our proof will follow the same strategy as Koras', using a momentum map for the action of the compact group G and an analytic continuation argument based on Weyl's unitary trick. But at the crucial point, instead of invoking the curve selection lemma, we will rely on an "interpolation" argument, deforming the Kähler metric in the neighbourhood of a fixed point.

Let us now explain the statement of the theorem and introduce some notation. We denote the complex structure on M by J, the Kähler metric by ds^2 and the Kähler form $-\operatorname{Im} ds^2$ by ω. Then $\operatorname{Re} ds^2 = \omega(\cdot, J\cdot)$ is the corresponding Riemannian metric. We may assume without loss of generality that ds^2 is invariant under the compact group G. So the transformations on M defined by G are holomorphic and they are isometries with respect to the Kähler metric. Saying that the action of G is Hamiltonian means that for all ξ in the Lie algebra $\mathfrak{g}$ of G the vector field ξ_M on M induced by ξ is Hamiltonian. In this case we have a *momentum map* Φ from M

to the dual $\mathfrak{g}^*$ of the Lie algebra of G with the property that

$$d\Phi^\xi = \iota_{\xi_M}\omega$$

for all ξ. Here Φ^ξ is the ξ-th component of Φ, defined by $\Phi^\xi(m) = \big(\Phi(m)\big)(\xi)$. Because its components are Hamiltonian functions, a momentum map is uniquely determined up to additive constants. After averaging with respect to the given action on M and the coadjoint action on $\mathfrak{g}^*$ we may assume that the map Φ is G-equivariant. It is easy to give sufficient conditions for the existence of a momentum map, e.g., the first Betti number of M is zero, or the Kähler form ω is exact. (See e.g. Guillemin and Sternberg [1990], Weinstein [1983].) More surprisingly, by a theorem of Frankel [1959] a momentum map always exists if the action has at least one fixed point and M is *compact*.

If M is $\mathbf{C}^n$ with the standard Hermitian structure and the standard symplectic form Ω, then a momentum map is given by the formula

$$\Phi^\xi_{\mathbf{C}^n}(v) = 1/2\,\Omega\big(\xi_{\mathbf{C}^n}(v), v\big), \tag{1}$$

where $\xi_{\mathbf{C}^n}$ denotes the image of $\xi \in \mathfrak{g}$ in the Lie algebra $\mathfrak{sp}(\mathbf{C}^n, \Omega)$, and $v \in \mathbf{C}^n$. In this setting the statement of the theorem is of course a tautology, but we will need (1) further on.

The decomposition of the complexified Lie algebra $\mathfrak{g}^\mathbf{C} = \mathfrak{g}\otimes\mathbf{C}$ into a direct sum $\mathfrak{g}^\mathbf{C} = \mathfrak{g} \oplus \sqrt{-1}\,\mathfrak{g}$ gives rise to the Cartan decomposition $G^\mathbf{C} = G\exp\sqrt{-1}\,\mathfrak{g}$ ("polar coordinates" on the group). The restriction of the exponential map $\exp\colon \sqrt{-1}\,\mathfrak{g} \to G^\mathbf{C}$ is a diffeomorphism onto its image, and every element g of $G^\mathbf{C}$ can be uniquely decomposed into a product $g = k\exp\sqrt{-1}\,\xi$, with $k \in G$ and $\xi \in \mathfrak{g}$.

Because $G^\mathbf{C}$ acts holomorphically on M, for any ξ in $\mathfrak{g}$ the vector field $(\sqrt{-1}\,\xi)_M$ induced by $\sqrt{-1}\,\xi$ is equal to $J\xi_M$. It follows easily from the definition of a momentum map that $J\xi_M$ is equal to the gradient vector field (with respect to the Riemannian metric $\mathrm{Re}\,ds^2$) of the ξ-th component of the momentum map,

$$(\sqrt{-1}\,\xi)_M = J\xi_M = \mathrm{grad}\,\Phi^\xi. \tag{2}$$

This elementary observation will enable us to gain control over the behaviour of the action "at infinity in the group". For one thing, it implies that the trajectory $\gamma(t)$ of $\mathrm{grad}\,\Phi^\xi$ through a point x in M is given by $\gamma(t) = \exp(\sqrt{-1}\,t\xi)x$, which does not depend on the choice of the Kähler metric or the momentum map.

Proof of Theorem 1. The proof is in three steps. In the first step we reduce the statement of the theorem to a statement about the trajectories of the gradient vector fields $(\sqrt{-1}\,\xi)_M = \mathrm{grad}\,\Phi^\xi$. In the second step we consider the case where the Kähler metric ds^2 is flat in some neighbourhood of m. The third step consists in showing that an arbitrary metric ds^2 can always be deformed to a metric which is flat close to m and which is still compatible with all the relevant data.

Step 1. The tangent space T_mM at m is a Hermitian vector space, which we shall identify with standard $\mathbf{C}^n$. Then the value of the Kähler form ω at M is the standard symplectic form Ω on $\mathbf{C}^n$. The tangent action of $G^\mathbf{C}$ defines a linear representation $G^\mathbf{C} \to \mathrm{GL}(n,\mathbf{C})$, the restriction of which to G is a unitary representation $G \to \mathrm{U}(n)$. Let $\phi\colon B \to \mathbf{C}^n$ be a local holomorphic coordinate system on M with $\phi(m) = 0$ and $d\phi(m) = \mathrm{id}_{\mathbf{C}^n}$, where B is a small G-invariant open neighbourhood of m such that $B' = \phi(B)$ is a ball about the origin in $\mathbf{C}^n$. Then the pullback of the form Ω is equal to ω at the point m. After averaging over G and shrinking B if necessary we may assume that ϕ is G-equivariant. Let $\psi\colon B' \to B$ be the inverse of ϕ.

We claim that if B is sufficiently small ϕ and ψ can be uniquely extended to holomorphic $G^{\mathbf{C}}$-equivariant maps $\phi^{\mathbf{C}}\colon U \to U'$ and $\psi^{\mathbf{C}}\colon U' \to U$, where $U = G^{\mathbf{C}}B$ and $U' = G^{\mathbf{C}}B'$. If this can be done, it is clear that $\phi^{\mathbf{C}} \circ \psi^{\mathbf{C}} = \mathrm{id}_{U'}$ and $\psi^{\mathbf{C}} \circ \phi^{\mathbf{C}} = \mathrm{id}_U$, so then the theorem is proved.

Following Heinzner [1991] we shall call an open G-invariant subset A of a $G^{\mathbf{C}}$-space X *orbitally convex* with respect to the $G^{\mathbf{C}}$-action if for all x in U and all ξ in $\mathfrak{g}$ the intersection of the curve $\{\exp(\sqrt{-1}\,t\xi)x : t \in \mathbf{R}\}$ with A is connected.

Proposition 2 (Heinzner [1991], Koras [1986]) *Let X and Y be complex manifolds acted upon by $G^{\mathbf{C}}$. If A is an orbitally convex open subset of X and $f\colon A \to Y$ is a G-equivariant holomorphic map, then f can be uniquely extended to a $G^{\mathbf{C}}$-equivariant holomorphic map $f^{\mathbf{C}}\colon G^{\mathbf{C}}A \to Y$.*

Consequently, if the image $f(A)$ is open and orbitally convex in Y and $f\colon A \to f(A)$ is biholomorphic, then the extension $f^{\mathbf{C}}\colon G^{\mathbf{C}}A \to Y$ is biholomorphic onto the open subset $G^{\mathbf{C}} \cdot f(A)$.

Proof: The only way to extend f equivariantly is by putting $f^{\mathbf{C}}\big(g\exp(\sqrt{-1}\,\xi)x\big) = g\exp(\sqrt{-1}\,\xi)f(x)$ for all x in A, g in G and ξ in $\mathfrak{g}$. We have to check this is well-defined.

Let $x \in A$ and $\xi \in \mathfrak{g}$ be such that $\exp(\sqrt{-1}\,\xi)x \in A$. Then by assumption $\exp(\sqrt{-1}\,t\xi)x \in A$ for all t between 0 and 1. So $f\big(\exp(\sqrt{-1}\,t\xi)x\big)$ is well-defined for $0 \le t \le 1$. Define the curves $\alpha(t)$ and $\beta(t)$ in Y by $\alpha(t) = f\big(\exp(\sqrt{-1}\,t\xi)x\big)$ and $\beta(t) = \exp(\sqrt{-1}\,t\xi)f(x)$ for $0 \le t \le 1$. Then $\alpha(t)$ and $\beta(t)$ are integral curves of the vector fields $f_*(\sqrt{-1}\,\xi)_X$ and $(\sqrt{-1}\,\xi)_Y$ respectively, both with the same initial value $f(x)$. Now since f is G-equivariant we have $f_*\xi_X = \xi_Y$, and, because f is also holomorphic, $f_*(\sqrt{-1}\,\xi)_X = f_*(J\xi_X) = Jf_*\xi_X = J\xi_Y = (\sqrt{-1}\,\xi)_Y$. Hence $\alpha(t) = \beta(t)$, in other words $f\big(\exp(\sqrt{-1}\,t\xi)x\big) = \exp(\sqrt{-1}\,t\xi)f(x)$ for $0 \le t \le 1$.

It follows that for all x in A and all ξ in $\mathfrak{g}$ such that $\exp(\sqrt{-1}\,\xi)x$ is in A we have $f\big(\exp(\sqrt{-1}\,\xi)x\big) = \exp(\sqrt{-1}\,\xi)f(x)$. It is easy to deduce from this that $f^{\mathbf{C}}$ is well-defined.

Finally observe that if the image $f(A)$ is open and orbitally convex in Y and $f\colon A \to f(A)$ is biholomorphic, then the inverse f^{-1} also has a holomorphic extension $(f^{-1})^{\mathbf{C}}\colon G^{\mathbf{C}}f(A) \to G^{\mathbf{C}}A$, and by uniqueness this must be the inverse of $f^{\mathbf{C}}$. $\square$

This proposition reduces the proof of Theorem 1 to showing that $B \subset M$ and $B' \subset \mathbf{C}^n$ are orbitally convex.

Step 2. We will now show that B is orbitally convex if the Kähler metric is flat on an open subset containing B. Since $\mathbf{C}^n$ is flat, this covers the case of $B' \subset \mathbf{C}^n$, so in particular we will have shown that B' is orbitally convex.

The argument is a variation on a convexity argument of Kempf and Ness [1979]. The assumption that the Kähler metric is flat on an open subset containing B implies that the holomorphic coordinate map $\phi\colon B \to \mathbf{C}^n$ is a Kähler isometry onto the open ball B' about the origin. After choosing appropriate constants we may assume that $\Phi(m) = 0$. Then $\psi^*\Phi\colon B' \to \mathfrak{g}$ is equal to the quadratic momentum map given by (1). (Recall $\psi = \phi^{-1}$.) In order not to overburden the notation we will in the proof of the following lemma identify B with B' and Φ with the momentum map (1). By $r(x)$ we denote the Riemannian distance of $x \in B$ to m.

Lemma 3 *Assume the Kähler metric is flat in the neighbourhood of B. Then for all $\xi \in \mathfrak{g}$ and $v \in B$ the momentum function Φ^ξ measures the inner product of the outward pointing normal* $\operatorname{grad} r^2$ *to the metric sphere of radius r about m and the vector field $J\xi_B = \operatorname{grad} \Phi^\xi$, as follows:*

$$\big\langle \operatorname{grad} r^2(v), \operatorname{grad} \Phi^\xi(v)\big\rangle = 4\Phi^\xi(v). \tag{3}$$

It follows that B is orbitally convex with respect to the $G^{\mathbf{C}}$-action.

Proof: The path $\delta(t) = \exp(\sqrt{-1}\,t\xi)v$ is the trajectory of the gradient of Φ^ξ through v. On one hand,

$$\frac{d}{dt} r^2\big(\delta(t)\big) = \big\langle \operatorname{grad} r^2\big(\delta(t)\big), \delta'(t)\big\rangle = \big\langle \operatorname{grad} r^2\big(\delta(t)\big), \operatorname{grad} \Phi^\xi\big(\delta(t)\big)\big\rangle.$$

On the other hand,

$$\begin{aligned}\frac{d}{dt} r^2\big(\delta(t)\big) &= \frac{d}{dt}\|\delta(t)\|^2 = \frac{d}{dt}\langle \delta(t), \delta(t)\rangle = 2\langle \delta'(t), \delta(t)\rangle = \\ &= 2\big\langle (\sqrt{-1}\,\xi)_B\big(\delta(t)\big), \delta(t)\big\rangle = 2\Omega\big(\xi_B\big(\delta(t)\big), \delta(t)\big) = 4\Phi^\xi\big(\delta(t)\big),\end{aligned}$$

where we have used (1) and (2). Taking $t = 0$ yields (3).

Now (3) implies that the curve $\delta(t)$ can only enter B at a point p in the boundary ∂B for which $\Phi^\xi(p) \leq 0$ and leave it at a point $q \in \partial B$ where $\Phi^\xi(q) \geq 0$. But $\delta(t)$ is also a gradient curve of the function Φ^ξ and so Φ^ξ is increasing along $\delta(t)$ for all $t \in \mathbf{R}$. If $\delta(t)$ is not constant, $\Phi^\xi\big(\delta(t)\big)$ is strictly increasing. Therefore, if $\delta(t)$ leaves the ball B at some point, it can never sneak back in. Consequently $\{\,\delta(t) : t \in \mathbf{R}\,\} \cap B$ is connected. If $\delta(t)$ is constant it is trivially true that $\{\,\delta(t) : t \in \mathbf{R}\,\} \cap B$ is connected. □

Step 3. We claim there exists a Kähler metric $d\tilde{s}^2$ on M with the following properties: It is compatible with the holomorphic structure J and G-equivariant, the G-action is Hamiltonian with respect to $d\tilde{s}^2$, and $d\tilde{s}^2$ is flat in a neighbourhood of m. Together with the result of step 2 this will conclude the proof of Theorem 1.

Let u be a Kähler potential for ds^2 in a neighbourhood of m, i.e.,

$$\omega = \sqrt{-1}\,\partial\bar{\partial} u.$$

Consider the Taylor expansion $u(x) \sim \sum_{k,l} c_{kl} x^k \bar{x}^l$ of u about m, where x is the system of holomorphic coordinates defined by the map $\phi\colon B \to \mathbf{C}^n$, and where k and l are multi-indices. The value of ω at m is the standard symplectic form on $\mathbf{C}^n$, so $\partial^2 u(m)/\partial x_\alpha \partial \bar{x}_\beta = \delta_{\alpha\beta}$. Also, deleting the holomorphic terms $\sum_k c_{k0} x^k$ and the antiholomorphic terms $\sum_l c_{0l} \bar{x}^l$ up to any finite order does not affect $\partial\bar{\partial} u$, so we may assume

$$u(x) = \|x\|^2 + \mathcal{R}(x),$$

with remainder term $\mathcal{R}(x)$ of order $O\big(\|x\|^3\big)$. Let $\chi\colon \mathbf{R} \to [0,1]$ be a smooth function with $\chi(t) = 0$ for $t \leq 1$ and $\chi(t) = 1$ for $t \geq 2$. Pick $\lambda > 0$ and put

$$\tilde{u}(x) = \|x\|^2 + \chi\big(\|x\|^2/\lambda^2\big)\mathcal{R}(x).$$

Let B_r be the ball $\{\, x \in B : \|x\|^2 < r\,\}$ and define a smooth two-form $\tilde{\omega}$ on M by

$$\tilde{\omega} = \begin{cases} \sqrt{-1}\,\partial\bar{\partial}\tilde{u} & \text{on } B_{3\lambda^2} \\ \omega & \text{on } M - B_{2\lambda^2}. \end{cases}$$

Then on $B_{2\lambda^2}$ the form $\tilde{\omega}$ is equal to the standard symplectic form Ω and so is G-invariant there. On $B_{3\lambda^2}$ we have $\tilde{\omega} - \omega = \sum_{\alpha,\beta=1}^{n} f_{\alpha\beta}\, dx_\alpha \wedge d\bar{x}_\beta$ with

$$f_{\alpha\beta}(x) = \frac{\partial^2}{\partial x_\alpha \partial \bar{x}_\beta}\Big(\big(\chi(\|x\|^2/\lambda^2) - 1\big)\mathcal{R}(x)\Big).$$

By carrying out the differentiation and using $\chi(t) = 1$ for $t \geq 2$ and $\mathcal{R}(x) = O\big(\|x\|^3\big)$ one can check in a straightforward manner that for every $\varepsilon > 0$ there exists $\Lambda > 0$ such that for all $\lambda < \Lambda$ and for all $x \in B_{3\lambda^2}$ one has $|f_{\alpha\beta}(x)| < \varepsilon$. In other words, $\tilde{\omega}$ can be made arbitrarily close to ω uniformly on M. It follows that for ε and λ small enough the symmetric bilinear form $\tilde{\omega}(\cdot, J\cdot)$ is positive-definite, and therefore $-\tilde{\omega}$ is the imaginary part of a Kähler metric $d\tilde{s}^2$. By construction $d\tilde{s}^2$ is equal to ds^2 on $M - B_{2\lambda^2}$ and flat on B_{λ^2}. After averaging over G we may assume $d\tilde{s}^2$ is G-invariant everywhere. (This does not destroy the flatness on B_{λ^2}, because $d\tilde{s}^2$ was already invariant there.)

It remains to be proved that the G-action is Hamiltonian with respect to $\tilde{\omega}$. By the Poincaré Lemma we can find an $\tilde{\omega}$-momentum map $\tilde{\Phi}\colon B_{3\lambda^2} \to \mathfrak{g}^*$. On the set $B_{3\lambda^2} - B_{2\lambda^2}$ we have $\tilde{\omega} = \omega$, so there $\tilde{\Phi}$ differs by a locally constant function c from the ω-momentum map Φ. Since $B_{3\lambda^2} - B_{2\lambda^2}$ is connected, c is a constant. Shifting Φ by this constant we can paste Φ and $\tilde{\Phi}$ together to obtain a global $\tilde{\omega}$-momentum map for the G-action. This finishes the proof of Theorem 1.

In Sjamaar [1995] we will show how to construct slices for a $G^{\mathbf{C}}$-action on a Kähler manifold at points m for which the real orbit Gm is *isotropic.* (This obviously includes the case of a fixed point.) Let us recall the definition of a slice.

Definition 4 *A* slice *at m for the $G^{\mathbf{C}}$-action is a locally closed analytic submanifold S of M with the following properties:*

1. *$m \in S$;*
2. *the saturation $G^{\mathbf{C}}S$ of S is open in M;*
3. *S is invariant under the action of the stabilizer $(G^{\mathbf{C}})_m$;*
4. *the natural $G^{\mathbf{C}}$-equivariant map from the associated bundle $G^{\mathbf{C}} \times_{(G^{\mathbf{C}})_m} S$ into M, which sends an equivalence class $[g, y]$ to the point gy, is an analytic isomorphism onto $G^{\mathbf{C}}S$.*

It follows from (4) that if a slice at m exists, then for all y in a $G^{\mathbf{C}}$-invariant neighbourhood of m the stabilizer $(G^{\mathbf{C}})_y$ is conjugate to a subgroup of $(G^{\mathbf{C}})_m$. Furthermore, a slice has to be transverse to the orbit $G^{\mathbf{C}}m$. Theorem 1 clearly shows that slices exist at fixed points, where $(G^{\mathbf{C}})_m$ is equal to $G^{\mathbf{C}}$.

The basic idea of the proof of the slice theorem in Sjamaar [1995] is the same as that of Theorem 1: We show that an isotropic G-orbit has sufficiently small orbitally convex neighbourhoods and then use analytic continuation. The main technical difficulty is to make the interpolation argument of step 3 work globally along an isotropic orbit. This involves a closer scrutiny of the second-order operator $\partial\bar{\partial}$.

The following example shows that slices need not exist everywhere, even at points m where the action is free, that is, $(G^{\mathbf{C}})_m = 1$.

Example 5 (Luna [1973]) Let V be the vector space of homogeneous cubic polynomials in two indeterminates x and y. Let G be the group SU(2); then $G^{\mathbf{C}} = \mathrm{SL}(2, \mathbf{C})$. Both groups act by linear substitutions in the two variables. If we

Table 1

	number of distinct factors	m	$(G^{\mathbf{C}})_m$
I	3	$(x+y)(x-y)y$	$\mathbf{Z}/3$
II	2	x^2y	1
III	1	y^3	$\mathbf{Z}/3$-extension of $\mathbf{C}$
IV	0	0	$G^{\mathbf{C}}$

declare the monomials x^3, x^2y, xy^2 and y^3 to be an orthonormal basis, we obtain a G-invariant Hermitian inner product on V. Every polynomial in V can be written as a product of three linear factors, which are unique up to scalar multiples. So, obviously, every element of $G^{\mathbf{C}}$ fixing the polynomial has to permute these factors (up to scalars). We can distinguish between four types of polynomials according to the number of distinct factors in the factorization, and it is easy to compute their stabilizers. (See table 1.) The points of type I form clearly a dense subset of V. Consider the point $m = x^2y$. The action is free at m, so the orbit $G^{\mathbf{C}}m$ is a copy of the three-dimensional $\mathrm{SL}(2,\mathbf{C})$ embedded in the four-dimensional V. A straightforward computation shows the orthogonal complement to $G^{\mathbf{C}}m$ at m is the one-dimensional subspace generated by y^3. We can define a holomorphic $G^{\mathbf{C}}$-equivariant map ϕ from the associated bundle $E = G^{\mathbf{C}} \times_{(G^{\mathbf{C}})_m} \mathbf{C} = \mathrm{SL}(2,\mathbf{C}) \times \mathbf{C}$ into V by putting $\phi\big([g,\eta]\big) = g \cdot (x^2y + \eta y^3)$. The differential of ϕ is bijective at all points of the zero section of E and the restriction of ϕ to the zero section is an embedding. Therefore ϕ is an open immersion in a neighbourhood of the zero section. Does there exist a $G^{\mathbf{C}}$-invariant neighbourhood U of the zero section of E such that the restriction of ϕ to U is biholomorphic onto its image? We could draw this conclusion if we knew that ϕ was injective on a $G^{\mathbf{C}}$-invariant neighbourhood of the zero section. But there are points of type I arbitrarily close to m, whose stabilizer consists of three elements and so cannot be conjugate to a subgroup of $(G^{\mathbf{C}})_m = 1$. Hence there cannot possibly exist a slice at m, and therefore the map ϕ cannot be injective in any $G^{\mathbf{C}}$-invariant neighbourhood of the zero section of E. The reason is that ϕ plays havoc with the fibres of E. Let us write $\eta = -\varepsilon^2$; then $\phi\big([g,\eta]\big) = gm_\varepsilon$ with $m_\varepsilon = (x+\varepsilon y)(x-\varepsilon y)y$. Let U be any $\mathrm{SL}(2,\mathbf{C})$-invariant neighbourhood of the zero section in E. Then U contains an invariant open neighbourhood of the form $E_\Delta = \mathrm{SL}(2,\mathbf{C}) \times \Delta$, where Δ is a small disc about the origin in $\mathbf{C}$. Choose ε such that $\varepsilon^2 \in \Delta$. The stabilizer $(G^{\mathbf{C}})_{m_\varepsilon}$ is generated by the following matrix of order three:

$$A_\varepsilon = \begin{pmatrix} -1/2 & 3\varepsilon/2 \\ -1/2\varepsilon & -1/2 \end{pmatrix},$$

and we have

$$A_\varepsilon m = -\frac{1}{8\varepsilon}x^3 + \frac{5}{8}x^2y - \frac{3\varepsilon}{8}xy^2 - \frac{9\varepsilon^2}{8}y^3.$$

So A_ε moves m way out along its orbit while leaving m_ε fixed. We conclude that the images of the fibres $\{I\} \times \Delta \subset E_\Delta$ and $\{A_\varepsilon\} \times \Delta \subset E_\Delta$ under the map ϕ intersect at the point m_ε.

Acknowledgement We are grateful to Eugenio Calabi for showing us how to glue Kähler potentials.

References

Bott, R. [1988], *Morse theory indomitable,* Inst. Hautes Études Sci. Publ. Math., **68**, 99–114.

Frankel, T. [1959], *Fixed points on Kähler manifolds,* Ann. of Math., (2) **70**, 1–8.

Guillemin, V., and Sternberg, S. [1982], *Geometric quantization and multiplicities of group representations,* Invent. Math., **67**, 515–538.

Guillemin, V., and Sternberg, S. [1990], *Symplectic techniques in physics,* Cambridge Univ. Press, Cambridge, second reprint with corrections.

Heinzner, P. [1991], *Geometric invariant theory on Stein spaces,* Math. Ann., **289**, 631–662.

Heinzner, P., and Loose, F. [1994], *Reduction of complex Hamiltonian G-spaces,* Geom. and Functional Anal., **4**, 288-297.

Kempf, G., and Ness, L. [1979], *The length of vectors in representation space,* Algebraic Geometry (Copenhagen, 1978) (K. Lønsted, ed.), Lecture Notes in Mathematics, **732**, Springer-Verlag, Berlin-Heidelberg-New York, 233–244.

Koras, M. [1986], *Linearization of reductive group actions,* Group Actions and Vector Fields (Vancouver, 1981) (J. B. Carrell, ed.), Lecture Notes in Mathematics, **956**, Springer-Verlag, Berlin-Heidelberg-New York, 92–98.

Luna, D. [1973], *Slices étales,* Sur les groupes algébriques, Mém. Soc. Math. France, **33**, 81–105.

Marsden, J., and Weinstein, A. [1974], *Reduction of symplectic manifolds with symmetry,* Rep. Math. Phys., **5**, 121–130.

Meyer, K. [1973], *Symmetries and integrals in mathematics,* Dynamical Systems (Univ. of Bahia, 1971) (M. M. Peixoto, ed.), Academic Press, New York.

Mumford, D., Fogarty, J., and Kirwan, F. [1994], *Geometric invariant theory,* third ed., Ergebnisse der Mathematik und ihrer Grenzgebiete, **34**, Springer-Verlag, Berlin-Heidelberg-New York.

Sjamaar, R. [1995], *Holomorphic slices, symplectic reduction and multiplicities of representations,* Ann. of Math. (2) **141**, 87-129.

Snow, D. M. [1982], *Reductive group actions on Stein spaces,* Math. Ann., **259**, 79–97.

Weinstein, A. [1983], *Lectures on symplectic manifolds,* CBMS Regional Conf. Series in Math., **29**, Amer. Math. Soc., third printing.

Fields Institute Communications
Volume 8, 1996

Atomic Spectroscopy, Periodic Orbits and Generic Two-Parameter Bifurcations

K. R. Meyer
Department of Mathematical Sciences
University of Cincinnati, Cincinnati, Ohio 45221

J. B. Delos
Permanent address:
Physics Department, College of William and Mary,
Williamsburg VA 23185
Current address:
Institute for Theoretical Atomic and Molecular Physics
Harvard-Smithsonian Center for Astrophysics
60 Garden Street, Cambridge, MA 02138

J.-M. Mao
Mathematics Department
Hong Kong University of Science and Technology
Clear Water Bay, Kowloon, Hong Kong

1 Physics

It is possible to measure in the laboratory the periods T_k or the actions S_k of classical periodic orbits of an atomic electron. This is a deep and complicated statement. Electrons obey laws of quantum mechanics, not classical mechanics, so we know that it is impossible to measure both the position and the velocity of an electron, and we do not imagine that any one electron actually moves in time along a definite path. Instead, the electron's behavior is described by the Schrödinger wave equation.

There is, however, a close relationship between the propagation of Schrödinger waves and the motion of a stream of particles along classical paths. First, the centroid of a well-localized wave packet moves in time according to classical equations of motion. A packet can be made to follow closely a classical periodic orbit; then the time required for the packet to return to its original location is close to the period of the classical orbit. Second, the change of phase of a Schrödinger wave as it propagates from one point to another is approximately equal to the action $\int \mathbf{p} \cdot d\mathbf{q}$ integrated along a classical path between these points. Such changes of phase produce interference effects which are observable in laboratory experiments.

1991 *Mathematics Subject Classification.* Primary: 34A47, 81Q20; Secondary: 70H05.

By computer calculation of classical orbits of electrons, it is possible to construct quantum wave functions, and we can use these wave functions to interpret and even to predict the results of measurements of the absorption spectra of atoms (Du and Delos [1988]). In this paper, we will not give any details about how this is done. We will just pretend that electrons are classical particles that obey Newton's laws. We will say that our friends (Holle et al. [1988]) found a way to place an atom in a magnetic field, and to kick an electron out of the atom such that they knew precisely how much kinetic energy the electron was given, but they could not tell in what direction it was kicked. They measured the time required for the electron to return to the atom (or the classical action around its orbit).

A sample measurement is shown in Figure 1. The large peaks mean that, after repeating the measurement many times, they often found electrons returning after acquiring approximately 1.5, 3.0 or 4.5 units of action, but they did not ever find electrons returning with (for example) 0.7 units of action. We call this set of observations the recurrence spectrum for the given kinetic energy.

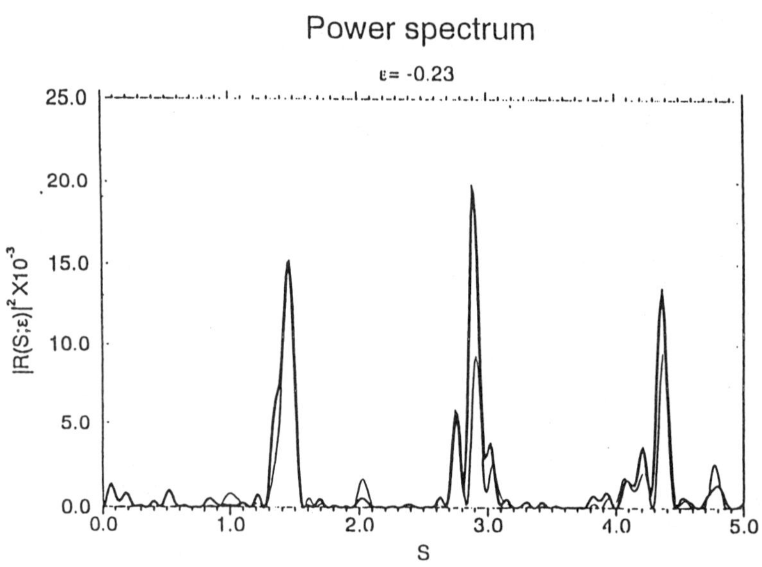

Figure 1 Heavy line: a measured recurrence spectrum for a hydrogen atom in a magnetic field. The horizontal coordinate represents classical action, and the vertical axis is called the recurrence strength; it is proportional to the number of electronics that return to the atom on orbits of given action. (It is actually measured by evaluating the Fourier transform of the absorption spectrum.) Light line: corresponding theoretical recurrence spectrum.

They then repeated the measurements with different initial kinetic energies given to the electron. The results are shown in Figure 2 – Main et al. [1994]. Again each peak corresponds to one or more recurrences – classical orbits which at the given energy return to the atom after acquiring a certain quantity of action. The most striking feature of this figure is the splitting and proliferation of peaks as the energy increases. These are bifurcations – as we vary the energy continuously, we pass through points at which a returning orbit splits, producing additional recurrences.

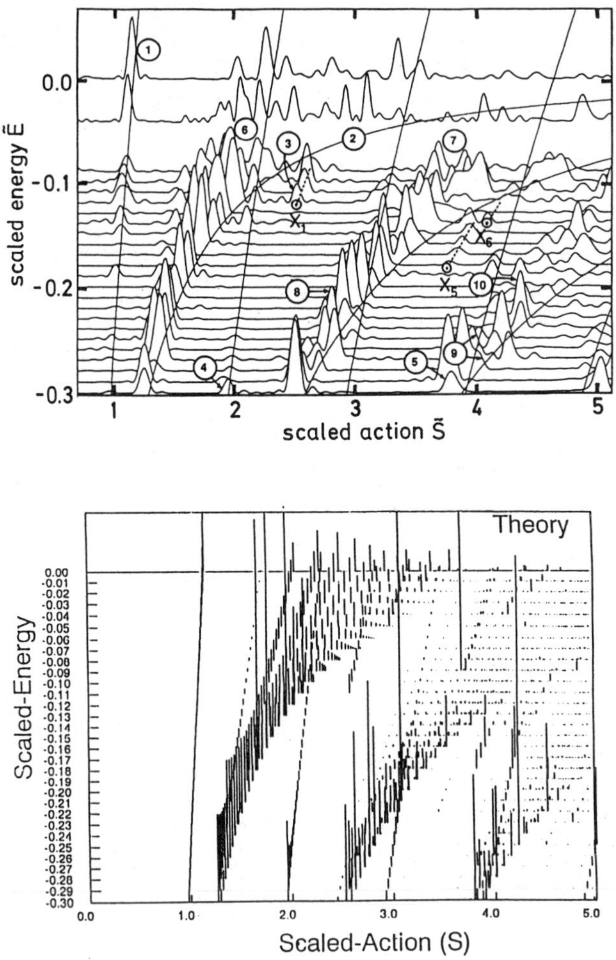

Figure 2 (a) About 25 recurrence spectra measured at various values of electron energy are drawn in a single graph. The horizontal axis is again classical action, while the vertical axis is electron energy, the parameter e in Hamiltonian H_1. (The electric field parameter f equals zero in this experiment.) (b) A corresponding family of theoretical recurrence spectra is shown as a needle graph. In both figures we see bifurcation and proliferation of recurrences with increasing e. This is correlated with the change from orderly to chaotic classical motion of the electron. The structures in this experiment are explained in detail by Main et al. [1994]. Peak 3 is produced in a saddle-node bifurcation, peaks 4 and 5 are associated with a 4-island chain bifurcation, the mountain range 6 is related to orbits that are produced in an orderly sequence of pitchforks and period-doublings, and peaks 9 and 10 arise from period-triplings.

Mathematical analysis shows that bifurcations of periodic orbits fall into certain characteristic patterns. Bifurcation theory is a very old subject, having roots in the work of Poincaré, but many long-neglected questions remain unanswered. A special aspect of the bifurcations visible here arises because atomic electrons obey Hamiltonian equations of motion, so their orbits conserve energy, and ensembles preserve volume in phase space and area in Poincaré planes. Bifurcations that occur in such systems have special structures that do not necessarily occur in more general, dissipative systems. Many macroscopic systems also obey Hamiltonian equations of motion to some degree of approximation, but nevertheless, it appears that these experiments are the first that *force* us to confront questions about the structure of bifurcations of periodic orbits in Hamiltonian systems.

With properly-chosen coordinates, the Hamiltonian for an atomic electron in fixed parallel electric and magnetic fields is

$$\begin{aligned} H_1 &= T(p_q,p_r)+V(q,r) \\ &= \frac{1}{2}(p_q^2+p_r^2)-4e(q^2+r^2) \\ &\quad -2q^2r^2(q^2+r^2)+2(q^6+r^6) \\ &\quad -16fqr(q^2+r^2) \end{aligned}$$

What is essential about this? (a) There are two degrees of freedom. (b) The kinetic energy is purely quadratic in the momenta, so the system has "time-reversal invariance". (c) The potential energy contains two parameters, e and f. (In atomic physics, the parameter e combines the conserved total energy and the magnetic field strength, while f is proportional to the electric field strength. Also, in this physical application, H_1 has the fixed value $H_1(p_q,p_r,q,r)=2$.) (d) The potential energy is invariant under inversion $(q,r)\to(-q,-r)$. (e) If $f=0$, the potential energy is invariant under reflections $(q,r)\to(-q,r)\to(q,-r)$. Small but nonzero values of f break this reflection symmetry. (Additional reflection symmetries are also present.) (f) The periodic orbits that are detected in the experiment are the ones that pass through the origin $(q{=}0,r{=}0)$ going in any direction.

Let us now pose general questions. (1) Given an arbitrary, smooth Hamiltonian function $H(p_q,p_r,q,r;e,f)$ defined in a phase-space (p_q,p_r,q,r) with smooth dependence on one or more parameters e,f, given that it possesses periodic orbits (PO's), how do these orbits bifurcate as the parameters are varied? What can be said about the geometrical structure of the newly created PO's? This question, fundamental to a physicist, may not be sufficiently well-posed to have any definite mathematical answer. However, a closely related question has been answered. (2) Such Hamiltonian differential equations induce two-dimensional area-preserving maps (Poincaré surfaces of section). What are the generic structures of bifurcations of PO's of smooth two-dimensional one-parameter area-preserving maps? This question was answered by Meyer [1970]; it was shown that there are five and only five generic types of bifurcation. (3) What structures are generic under symmetries, such as those mentioned above? This question was answered in Rimmer [1982], also see Mao and Delos [1992], deAguiar et al. [1987]. (4) What are the generic structures of two-dimensional two-parameter area-preserving maps?

To illustrate the importance of this question, let us consider a specific example. For $f=0$, the Hamiltonian H_1 has a periodic orbit having $r=0, p_r=0$; if we vary the parameter e, then at a certain point that orbit undergoes an "antipitchfork" bifurcation: it changes from stable to unstable, and simultaneously two other unstable PO's merge with it and are destroyed. This antipitchfork bifurcation is a nongeneric structure – it exists because of the symmetries of the system. It is common for an unstable PO to collide with a stable PO (at which point they annihilate each other), but it is not common for two distinct unstable PO's to collide simultaneously with a stable PO. Now if f is changed to a nonzero value, the reflection symmetry of $V(q, r)$ is broken and the stable PO moves off the $r = 0$ line. What happens when it bifurcates? Does it still change from stable to unstable? Is the antipitchfork structure still present? This change of f corresponds to imposing an electric field on the atom. A change in the structure of this bifurcation might have observable physical consequences for the system. What are they?

Full answers to all of these questions are not now available. We can begin below with a partial answer to question (4) above. We will show that pitchfork, antipitchfork, and period-doubling bifurcations break in certain characteristic ways when the symmetry is broken (Figure 3).

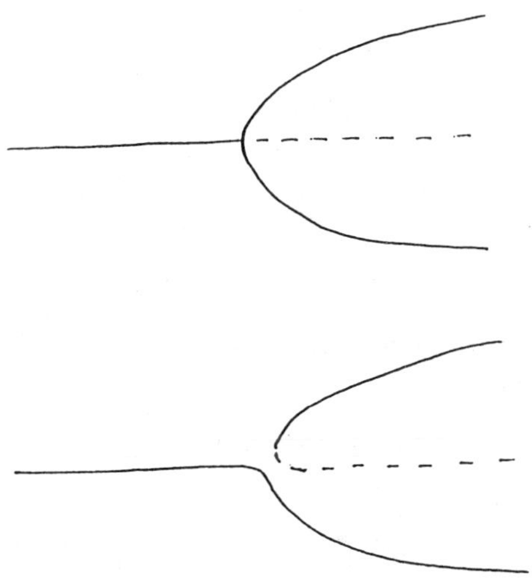

Figure 3 A representation of a pitchfork bifurcation (stable goes unstable and two new stable orbits of the same period are created), and a representation of a "broken pitchfork" (the stable orbit moves aside to make way for a saddle-node bifurcation

To emphasize that the result is very general, and does not depend in any way on the special physics of Hamiltonian H_1, we change the notation. The Poincaré plane is denoted (x, y) and we begin with an area-preserving mapping of the plane that depends on two parameters (μ, ν).

2 Mathematics

Notation. Let P be a smooth, area preserving mapping of a region $\mathbf{O}$ in the plane, $\mathbf{R}^2$, into the plane which depends on two parameters. Let $u = (x, y)$ be coordinates in the plane and let the two parameters be denoted by μ and ν. We are interested in the nature and bifurcation of fixed points and periodic points of P as the parameters are varied. The region $\mathbf{O}$ should contain an open neighborhood of the fixed point in question. We denote the two components of P by (X, Y) and so $P : (x, y) \to (X, Y)$. Sometimes we will show the dependence on the parameters by writing $X(x, y, \mu, \nu)$ etc. Iterates of P will be denoted by $P^k = P^\circ P^\circ \ldots^\circ P = (X^k, Y^k)$. There are several places where $\pm$ are needed, so α and β will be ± 1.

That P is area preserving means that the Jacobian matrix $DP(u) = \partial(X, Y)/\partial(x, y)$ has determinant identically equal to one. Let P have a fixed point at u_0, $P(u_0) = u_0$ and let $A = DP(u_0)$. A is a 2×2 matrix with determinant $+1$ and characteristic equation $\lambda^2 - (\text{trace } A)\lambda + 1$. Its eigenvalues are called the *(characteristic) multipliers* of the fixed point and since $\det A = 1$ they are reciprocals, thus the multipliers will be denoted by λ, λ^{-1}. The linearization or first approximation of P at the fixed point is the map $u \to Au$. Based on the linear approximation there are three basic cases:

1. Hyperbolic case: $|\text{trace } A| > 2$. The eigenvalues of A are real and satisfy $0 < \lambda < 1 < \lambda^{-1}$ or $\lambda^{-1} < -1 < \lambda < 0$. The fixed point is unstable by the linear approximation and by classical results the fixed point is still unstable when the higher order terms are included.

2. Elliptic case: $|\text{trace } A| < 2$. the eigenvalues of A are complex and lie on the unit circle so are of the form $\lambda = e^{i\theta}$, $\lambda^{-1} = e^{-i\theta}$. The linearized map is a rotation by the angle θ and so the origin is stable for the linear approximation. The higher order terms can make the origin unstable, but by the KAM theory except for some low resonance cases where λ is a small root of unity the origin is stable generically. See Moser [1962], [1968].

3. Parabolic case: $|\text{trace } A| = 2$. The eigenvalues of A are $+1, +1$ or $-1, -1$. The bifurcations and stability depends heavily on the higher order terms. We shall look at this interesting case in detail when additional generic assumptions are made on the higher order terms.

The multipliers tell a lot about the dependence and bifurcation of the fixed points. The first important observation comes from the implicit function theorem. A fixed point of P satisfies the equation

$$P(u, \mu, \nu) - u = 0. \tag{2.1}$$

Assume that $u = u_0$ is a fixed point for P for some value of the parameters, say $\mu = \mu_0$, $\nu = \nu_0$, thus (u_0, μ_0, ν_0) satisfies equation (2.1). The Jacobian of (2.1) at u_0 is $DP(u_0, \mu_0, \nu_0) - I = A - I$ and so if $+1$ is not a multiplier the hypotheses of the implicit function theorem hold and there is a smooth function $\eta(\mu, \nu)$ defined for μ, ν near μ_0, ν_0 which satisfies $P(\eta(\mu, \nu), \mu, \nu) = \eta(\mu, \nu)$. Moreover, $\eta(\mu, \nu)$ is the only solution near u_0. Thus the fixed point can be continued as an isolated fixed point for values of the parameters near μ_0, ν_0. The eigenvalues vary continuously with the parameters also. Thus a hyperbolic (respectively elliptic) continues as a hyperbolic (respectively elliptic) fixed point, but parabolic fixed points can and in general do change type.

The fixed point is called *elementary* if its multipliers are not $+1$. In this case we can change coordinates and parameters by $u' = u - \eta(\mu, \nu)$, $\mu' = \mu - \mu_0$, $\nu' = \nu - \nu_0$ so that in the primed variables the fixed point of P is at the origin for all small values of the parameters. We shall assume that this change has been made and drop the primes on the variables.

The above argument shows that a fixed point can bifurcate only when its multipliers are $+1$. However, there may be periodic points which bifurcate from the fixed point. The above argument shows that the fixed point is isolated as a periodic point of period k if $DP^k(0) = A^k$ does not have eigenvalues $+1$. This can happen only if the multipliers are roots of unity.

If there are no parameters then generically all the fixed points are either hyperbolic or elliptic and the multipliers of the elliptic points are not roots of unity. If there is one parameter then fixed points with multipliers ± 1 or other roots of unity occur generically for isolated values of the parameter. The bifurcations which can occur in generic one parameter families are described in Meyer [1970], but see the pictures in Abraham and Marsden [1978].

This paper will begin the classification of the generic bifurcation when the mapping P depends on two parameters. In this case there are curves in parameter space where P has fixed points with multipliers that are a specific root of unity. Only the parabolic case is considered here when the multipliers are ± 1. The case when the multipliers are $+1$ is discussed in the next section and when they are -1 in the subsequent section. The bifurcations for higher roots of unity will be the topic of a later paper.

Creation – Multipliers = +1. Poincaré [1899] associated to the map P a generating function G with the property that fixed points of P are critical points of G and vice versa. Specifically, let $P = (X, Y)$ be defined in a disk about the origin in the plane and consider the differential form

$$\Omega = (Y - y)d(X + x) - (X - x)d(Y + y). \tag{2.2}$$

Since P is area preserving $d(XdY - xdy) = dX \wedge dY - dx \wedge dy = 0$ and similarly $d(YdX - ydx) = dY \wedge dX - dy \wedge dx = 0$. These identities and

$$\Omega = (YdX - ydx) - (XdY - xdy) + d(xY - xX)$$

imply that $d\Omega = 0$, i.e. Ω is closed. Since we are working on a disk Poincaré's lemma asserts that a closed form is exact so there is a function G such that $dG = \Omega$.

The basic facts about this generating function are given in the following lemma and the proof can be found in Meyer [1970]. Also see the original discussion in Poincaré [1899] and another discussion in Weinstein [1972].

Lemma 2.1 *Assume that for $\mu = \nu = 0$ the map P has a fixed point at the origin whose multipliers are not equal to -1. Then the generating function G defined by (2.1) is a function of x, y for x, y, μ, ν small. It has the following properties:*

1. *A critical point of G (i.e. where $G_x = G_y = 0$) is a fixed point of P and conversely.*
2. *A nondegenerate maximum or minimum of G corresponds to an elliptic fixed point of P and conversely.*
3. *A nondegenerate saddle point of G corresponds to an hyperbolic fixed point of P and conversely.*

To obtain a rough picture of the map draw the level lines of G – but remember this is only a first approximation of the full picture.

From catastrophe theory if there are no parameters then generically G has only non-degenerate critical points and so generically P has only elliptic of hyperbolic fixed points. If there is one parameter then G can have fold critical points i.e. G might be of the form

$$G = \pm y^2/2 + x^3/6 - \mu x. \tag{2.3}$$

This gives rise to the elliptic-hyperbolic (sometimes called saddle-node) bifurcation discussed in Meyer [1970] and subsequent papers. Two parameters gives rise to the cusp catastrophe with generating function

$$G = \pm y^2/2 + x^4/12 - \mu x^2/2 + (2/3)\nu x = \alpha y^2/2 + x^4/12 - \mu x^2/2 + (2/3)\nu x. \tag{2.4}$$

See Poston and Stewart [1978] for a readable introduction to catastrophe theory and additional references.

Remark: *By Thom's theorem on the classification of elementary catastrophes there are coordinates such that a cusp catastrophe is of the above polynomial form, but these coordinates are not symplectic coordinates! Therefore, we will not derive the map from G, but simply study its critical points.*

The critical points of G are solutions of

$$G_y = \alpha y = 0, \qquad G_x = x^3/3 - \mu x + 2\nu/3 = 0 \tag{2.5}$$

The singular locus where G has a degenerate critical point is found by solving $G_x = 0$ and $G_{xx} = x^2 - \mu = 0$. This locus is the cusp $\mu^3 = \nu^2$. Figure 4 labels various regions in the parameter plane. Regions b and e are on the singular locus. Figure 5 shows the graphs of $g(x) = G(x, 0, \mu, \nu)$ as a function of x for the various regions in the parameter plane.

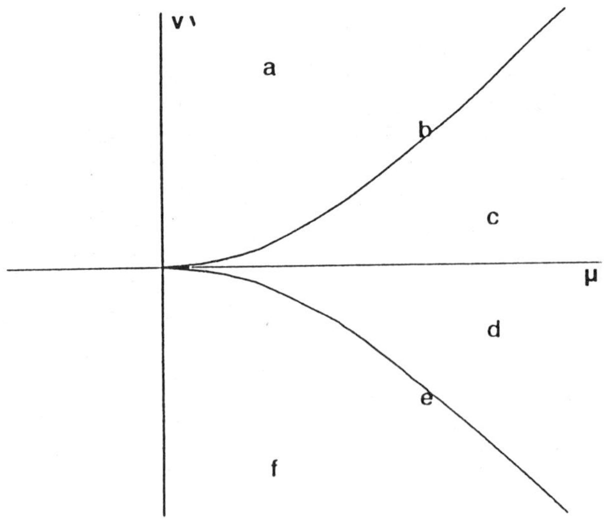

Figure 4 Regions in the parameter plane for the cusp bifurcation

If $\alpha = +1$ then a non-degenerate minimum x_0 of $g(x)$ gives rise to a non-degenerate minimum of G at $(x_0, 0)$ and hence to an elliptic fixed point of the map

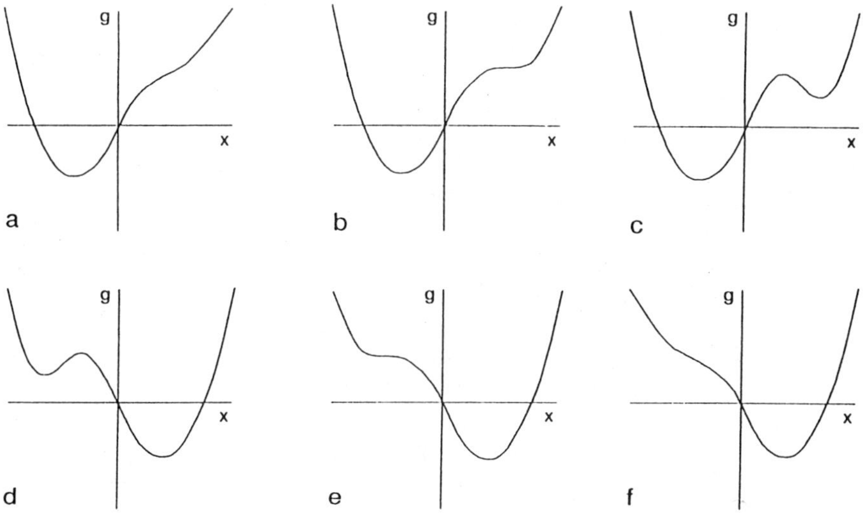

Figure 5 Graphs of g vs x for various values of parameters.

P. If $\alpha = +1$ then a non-degenerate maximum x_0 of $g(x)$ gives rise to a non-degenerate saddle point of G at $(x_0, 0)$ and hence to a hyperbolic fixed point of the map P. See Figure 6.

When $\alpha = -1$ it is the other way around. Namely, if $\alpha = -1$ then a non-degenerate minimum x_0 of $g(x)$ gives rise to a non-degenerate saddle of G at $(x_0, 0)$ and hence to an hyperbolic fixed point of the map P and a non-degenerate maximum x_0 of $g(x)$ gives rise to a non-degenerate maximum of G at $(x_0, 0)$ and hence to an elliptic fixed point of the map P. See Figure 4.

Figures 4-7 summarize the new bifurcations which occur with two parameters when the fixed point has multiplier $+1$.

Period Doubling – Multipliers = -1. The discussion of period doubling for area preserving maps given below is a substantial improvement to the original discussion in Meyer [1970]. In this section we will proceed formally and not worry about the convergence of the normal form. One only has to fuss with the implicit function theorem to show that formal periodic points are true periodic points. In fact by Artin's implicit function theorem a formal analysis suffices in the analytic case – see Artin [1968].

Assume when $\mu = \nu = 0$ that P has a fixed point at the origin with multipliers -1. Then the Jacobian matrix of P at the origin is symplectically similar to one of the following

$$\begin{pmatrix} -1 & 0 \\ 0 & -1 \end{pmatrix} \qquad \begin{pmatrix} -1 & \pm 1 \\ 0 & -1 \end{pmatrix}, \tag{2.6}$$

see Meyer and Hall [1992]. The first matrix is so degenerate that it takes three parameters to unfold it – see Arnold [1972]. That is, generically the first matrix in (2.1) may appear in a three parameter family but not in a two parameter family. Since we are considering only the two parameter case we need only consider the second matrix.

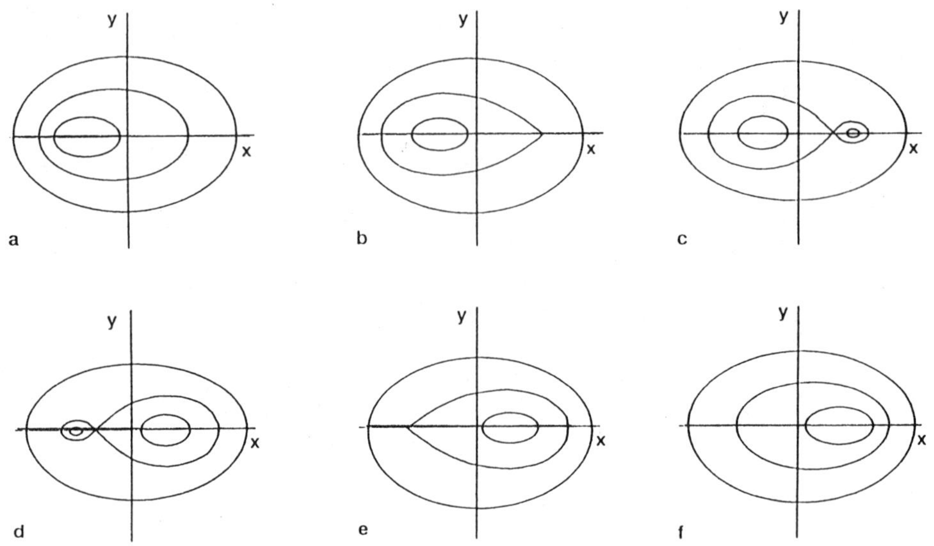

Figure 6 The cusp bifurcation, $\alpha = +1$.

Lemma 2.2 *Assume P has a fixed point at the origin with multipliers -1 and that $DP(0)$ is* $\begin{pmatrix} -1 & \alpha \\ 0 & -1 \end{pmatrix}$ *then the formal norm for P is*

$$P(u) = -\Phi(u) \tag{2.7}$$

where Φ is the time 1-map of a Hamiltonian system with Hamiltonian G where G is of the form

$$\begin{aligned} G(u) = G(x,y) &= \alpha y^2/2 + K(x) \\ &= \alpha y^2/2 + ax^2/2 + bx^4/4 + cx^6/6 + \cdots . \end{aligned} \tag{2.8}$$

Here as above $\alpha = \pm 1, a, b, c$ etc. are the constants (which depend on the parameters μ and ν of course). The formal series K is even.

Proof: See Meyer and Hall [1992] pages 197-198.

We shall proceed with a formal analysis, i.e. assuming the system is in normal form. It should be clear that our conclusions are based on only a finite number of terms in the normal form and so the system need only be in normal form up to a certain fixed degree. A liberal dose of the implicit function theorem will show that the formal analysis suffices, see for example the arguments in Meyer [1970] or in Hale and Chow [1982].

Lemma 2.3 *P has $u_0 = (x_0, y_0)$ as a periodic point of period 2 if $y_0 = 0$ and x_0 is a critical point of K $(K_x(x_0) = 0)$. Moreover, $u_0 = (x_0, y_0)$ is an elliptic point of P if and only if $u_0 = (x_0, 0)$ is a non-degenerate maximum or minimum of G and $u_0 = (x_0, 0)$ is an hyperbolic point of P if and only if $u_0 = (x_0, 0)$ is a non-degenerate saddle point of G.*

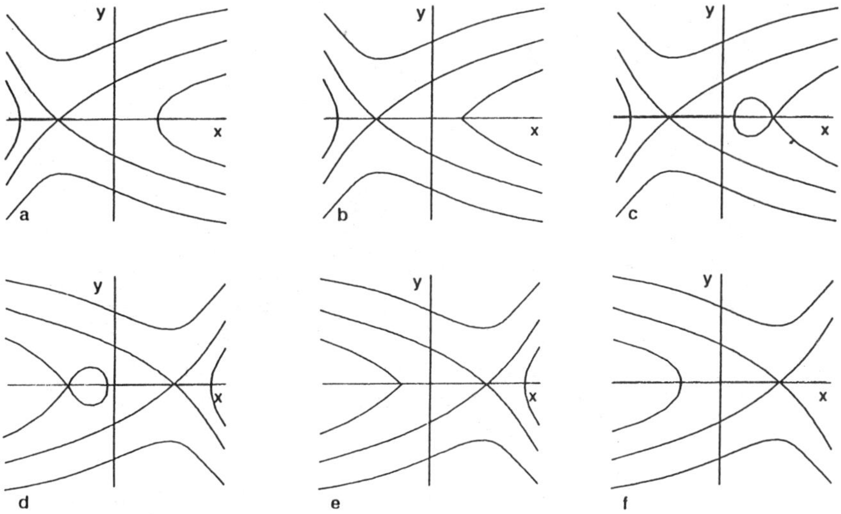

Figure 7 The cusp bifurcation, $\alpha = -1$.

Proof: Since K is even, if $u_0 = (x_0, 0)$ is a critical point of G then so is $-u_0 = (-x_0, 0)$. Thus the Hamiltonian flow defined by G leaves $\pm u_0$ fixed, $\Phi(\pm u_0) = \pm u_0$. So $P(\pm u_0) = \mp u_0$ and u_0 is a periodic point of period 2.

Conversely, let u_0 be a 2-periodic point of P. Since K is even the time one map satisfies $\Phi(-u) = -\Phi(u)$, so $\Phi^2(u_0) = u_0$. This means that u_0 is on a periodic orbit of period 2 of the flow defined by G. Since this flow is a small perturbation of the flow defined by the Hamiltonian $\alpha y^2/2$ all non-critical point periodic orbits have very long periods. Thus u_0 is a critical point.

If $\pm u_0$ is a nondegenerate maximum or minimum of G then u_0 is an elliptic critical point of the flow defined by G and $A = JDG(\pm u_0)$ has pure imaginary eigenvalues. Thus the eigenvalues of $\exp(A)$ are on the unit circle as are those of $\exp(2A)$. But $\exp(2A) = DP^2(\pm u_0)$, so the fixed point is elliptic. A similar argument shows that a nondegenerate saddle point of G corresponds to a hyperbolic periodic point of P. The converse follows by the same line of reasoning. □

Thus all we need to do is study the generic unfolding of critical points of functions like G and really just K. We will call G the *generating function.*

In Meyer [1970] the condition for a generic period doubling bifurcation was complicated by the fact that the map was not put into a normal form. Then there was one parameter, μ and $a(\mu)$, $b(\mu)$, $c(\mu)$, etc. depended on the parameter. For the one parameter case the generic assumptions are $\alpha = \pm 1$, $a(0) = 0$, $a'(0) \neq 0$ and $b(0) \neq 0$.

Here is how to interpret these conditions: (1) $\alpha = \pm 1$ means that when the map has the multiplier -1 generically the highly degenerate first matrix in (6) can be avoided; (2) $a(0) = 0$ means that map has the multiplier -1 when $\mu = 0$; (3) $a'(0) \neq 0$ means the Jacobian of P passes through the second matrix in (6) with non-zero speed and multipliers are $\pm\sqrt{(-a\alpha\mu)} + \cdots$; (4) $b(0) \neq 0$ means the first term that cannot be made to vanish by a normalization transformation does not vanish.

With two parameters μ, ν one of the last two conditions can be violated for some specific value of the parameters. In particular we shall continue to assume $\alpha = \pm 1$, $a(0,0) = 0$. But we may have $a_\mu(0,0) = 0$ (case 1) or $b(0,0) = 0$ (case 2). In both these cases higher derivatives are generically nonzero.

$$\textit{Case 1:} \quad \alpha = \pm 1, \; a(0,0) = 0, \; a_\mu(0,0) = 0, \; a_{\mu\mu}(0,0) \neq 0, \\ a_\nu(0,0) \neq 0, \; b(0,0) \neq 0.$$

The full normal form contains terms of degree six and higher, but these terms will not change the qualitative nature of the bifurcations. So we will truncate the series after the fourth order term. By Lemma 2.3 the placement and type of the periodic points do not depend on the sign of G, so by changing the sign of G if necessary the coefficient b can be made positive. Now scale the variable x so that the coefficient of x^4 is 1/4. The values of α and a may change but not the form of G.

By assumption a is of the form $a(\mu, \nu) = d(\mu, \nu)\mu^2 + 2e(\mu, \nu))\mu\nu + f(\mu, \nu)\nu$ where $d(0,0) \neq 0$ and $f(0,0) \neq 0$. Completing the square gives

$$a(\mu, \nu) = d\{\mu + [e/d]\nu\}^2 + \{f - [e^2/d]\nu)\nu\}.$$

Change parameters by $\mu' = \sqrt{|d|}\{\mu + [e/d]\nu\}$ and $\nu' = -\{f - [e^2/d]\nu\}\nu$. With these parameters $a(\mu', \nu') = \beta\mu'^2 - \nu'$ where $\beta = \pm 1$. Make this change of parameters and drop the primes.

The prototype generating function in this case is

$$G = \alpha y^2/2 + K(x), \;\; K(x) = -(\nu - \beta\mu^2)x^2/2 + x^4/4. \tag{2.9}$$

The critical points of G are at the origin and at $(x, 0)$ where x satisfies

$$\nu - \beta\mu^2 = x^2. \tag{2.10}$$

This equation has two solutions (one periodic orbit of period 2) when $\nu - \beta\mu^2$ is positive, namely at

$$x = \pm\sqrt{(\nu - \beta\mu^2)}, \tag{2.11}$$

and these are nondegenerate minima of K.

Since $G = \alpha y^2/2 + K(x)$ when $\alpha = +1$ a minimum of K gives rise to minimum of G and so to an elliptic point of P and a maximum of K gives rise to a saddle point of G and so to a hyperbolic point of P. When $\alpha = -1$ a minimum of K gives rise to a saddle point of G and so a hyperbolic point of P and a maximum of K gives rise to a maximum of G and so to an elliptic point of P.

Lemma 2.4 *When $\alpha = +1, P$ has an elliptic periodic point of period 2 which bifurcates from the origin when $\nu - \beta\mu^2$ is positive. When $\alpha = -1, P$ has an hyperbolic periodic point of period 2 which bifurcates from the origin when $\nu - \beta\mu^2$ is positive.*

The exact nature of the bifurcation will be illustrated in the following sequence of figures. We can rewrite (2.10) as

$$\nu = \beta\mu^2 + x^2 \tag{2.12}$$

and look at the equation as the equation for a conic section in the μ, x plane. In the case $\beta = +1$ then (2.12) is the equation of a circle for $\nu > 0$. In the case $\beta = -1$ then (2.12) is two lines through the origin when $\nu = 0$ and it is the equation of a hyperbola when ν is nonzero. Think of the nature of the bifurcations which occur as μ is varied for fixed ν. This is illustrated in the sequences of images in Figures 8 and 9. These figures seek to illustrate the full four dimensional phase-parameter space (x, y, μ, ν). Since $y = 0$ always it will be ignored and a sequence of x, μ planes are given for different values of ν.

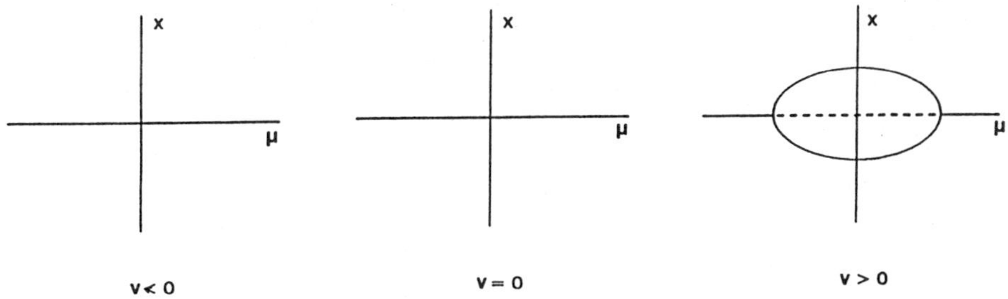

Figure 8 Period doubling in case 1 when $\beta > 0$.

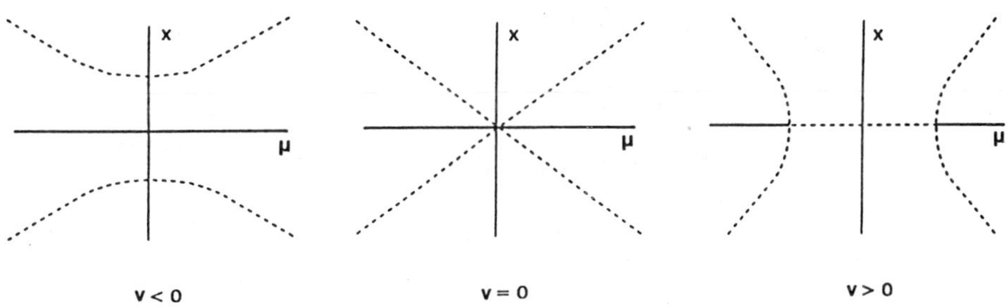

Figure 9 Period doubling in case 1 when $\beta < 0$.

$x = 0$ is always a fixed point and it is hyperbolic when $\alpha(\nu - \beta\mu^2)$ is positive and elliptic when it is negative. Thus when $\beta\nu < 0$ the origin is of the same type for all μ. But when $\beta\nu > 0$ the origin changes its stability type at $\mu = \pm\sqrt{\beta\nu}$. In the Figures solid lines and dashed lines represent periodic points of different stability type. A bifurcation occurs when the origin changes stability type.

Figure 8 illustrates $\beta = +1$ and Figure 9 illustrates $\beta = -1$. The origin is of one stability type for all μ when $\beta\nu < 0$. When $\beta\nu > 0$ the origin has the opposite stability type for $-\sqrt{\beta\nu} < \mu < \sqrt{\beta\nu}$. The periodic point bifurcates from the origin at $\pm\sqrt{\beta\nu}$ as illustrated.

To summarize this case, if in the normal form (3), $a_\mu(\mu,\nu) = 0$ for $\mu = \nu = 0$, then for $\nu < 0$, as μ increases we find an ordinary period-doubling bifurcation which then reverses itself. If $\beta = +1$ a stable period-1 orbit goes unstable and a new stable orbit of twice the period bifurcates out of it; the new orbit moves away, then comes back to the now-unstable period-1 orbit, collides with it, and disappears, leaving the original orbit stable again. (This phenomenon was seen in a (non-Hamiltonian) electrical network by Hilborn [1990], who dubbed it "period-bubbling".) If $\beta = -1$, as μ increases, an unstable orbit of period two approaches a stable period-1 orbit, collides with it and disappears, leaving the period-1 orbit unstable; then later the unstable period-2 orbit reappears, at which point the period-1 orbit goes stable. For either sign of β, the words "stable" and "unstable" can be interchanged in the above description.

$$\textit{Case 2:}\quad \alpha = \pm 1,\ \ a(0,0) = b(0,0) = 0\ \ c(0,0) \neq 0,$$
$$a_\mu(0,0)b_\nu(0,0) - a_\nu(0,0)b_\mu(0,0) \neq 0.$$

Again truncate K in Lemma 2.3 at the sixth order since higher order terms will not affect the qualitative features of the bifurcations and let the coefficient $c(\mu,\nu)$ be constant for simplicity. These assumptions simply make the analysis easier. Scale by $x \to |c|^{-1/6}x$ so that the coefficient x^6 in K is $\beta = \pm 1$.

Without loss of generality we can assume that $a \equiv \mu$ and $b \equiv \beta\nu$ since the Jacobian condition $a_\mu(0,0)b_\nu(0,0) - a_\nu(0,0)b_\mu(0,0) \neq 0$ implies that $\mu' = a(\mu,\nu)$, $\nu' = \beta b(\mu,\nu)$ is a valid change of parameters. Having made the change of parameters $(\mu,\nu) \to (\mu',\nu')$ we drop the primes.

Thus in this case the generating function will be

$$G = \alpha y^2/2 + \beta K(x),\ \ K(x) = \beta\mu x^2/2 + \nu x^4/4 + x^6/6. \tag{2.13}$$

The critical points of G are at $y = 0$ and x_0 where x_0 is a critical point of K. The critical points of K are at $x = 0$ and at solutions of

$$x^4 + \nu x^2 + \beta\mu = 0. \tag{2.14}$$

Equation (2.14) is a quadratic in x^2 so easy to solve. Roots are symmetric about the origin since (2.14) is even. Let the set of roots be $\mathcal{R}$. The following summarizes the real solutions of (2.14) (the regions refer to regions in parameter space shown in Figure 10):

Since $G = \alpha y^2/2 + \beta K(x)$ when α and β are the same sign a minimum of K gives rise to minimum or maximum of G and so to an elliptic point of P and a maximum of K gives rise to saddle point of G and so to a hyperbolic fixed point of P. When α and β are of opposite sign a minimum of K gives rise to a saddle point of G and so a hyperbolic point of P a maximum of K gives rise to minimum or maximum of G and so to an elliptic point of P.

Figure 11 shows the graph of K in the various regions, and Figures 12 and 13 show the fixed points of P in the various regions.

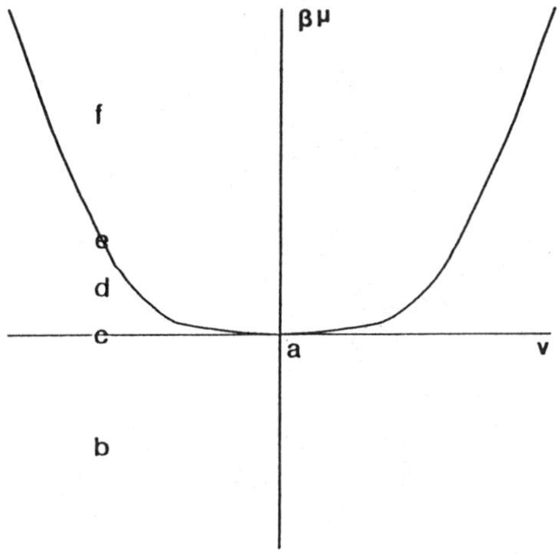

Figure 10 Parameter space for period doubling, case 2.

$$
\begin{aligned}
\mathcal{R} &= \{0,0,0,0\} \text{ when } \mu=\nu=0 \text{ (Region a).}\\
\mathcal{R} &= \{x_1,-x_1\}, x_1>0 \text{ when } \beta\mu<0. \text{ (Region b).}\\
\mathcal{R} &= \{x_1,0,-x_1,0\}, x_1>0 \text{ when } \mu=0 \text{ and } \nu<0 \text{ (Region (c).}\\
\mathcal{R} &= \{x_1,x_2,-x_1,-x_2\}x_1>x_2>0 \text{ when } 0<4\beta\mu<\nu^2 \text{ (Region d).}\\
\mathcal{R} &= \{x_1,x_1,-x_1,-x_1\}, x_1>0 \text{ when } 4\beta\mu=\nu^2 \text{ (Region e).}\\
\mathcal{R} &\quad \text{is empty otherwise (Region f).}
\end{aligned}
$$

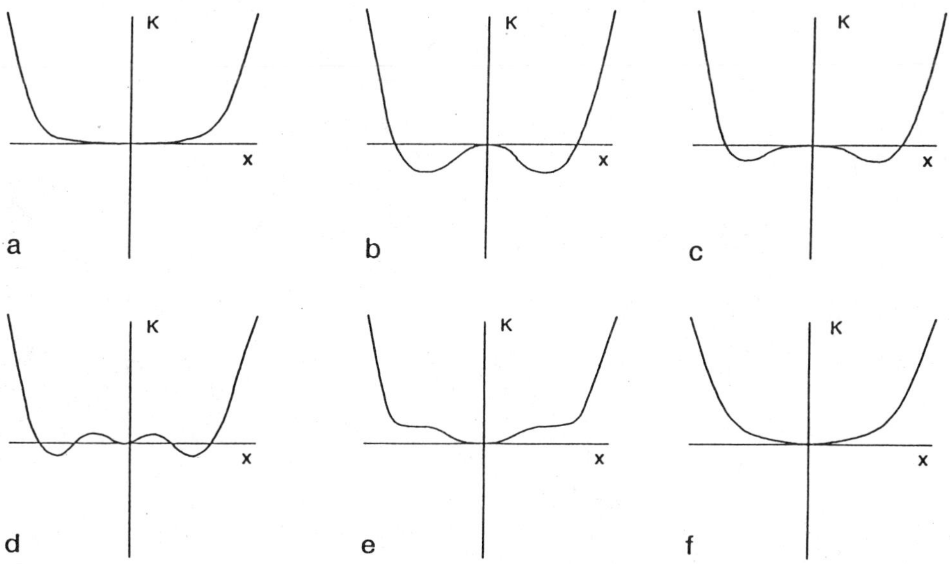

Figure 11 Period doubling case 2, K vs x.

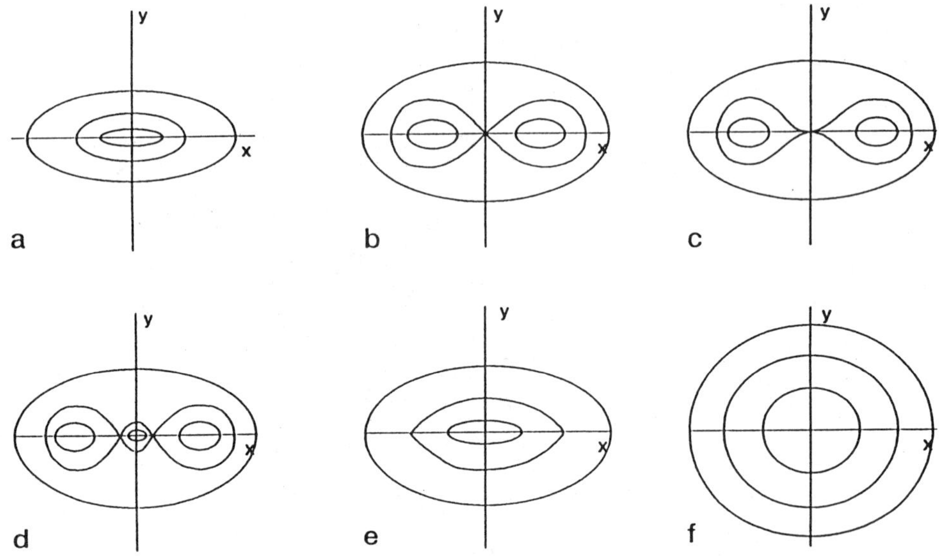

Figure 12 Period doubling case 2, $\alpha\beta > 0$.

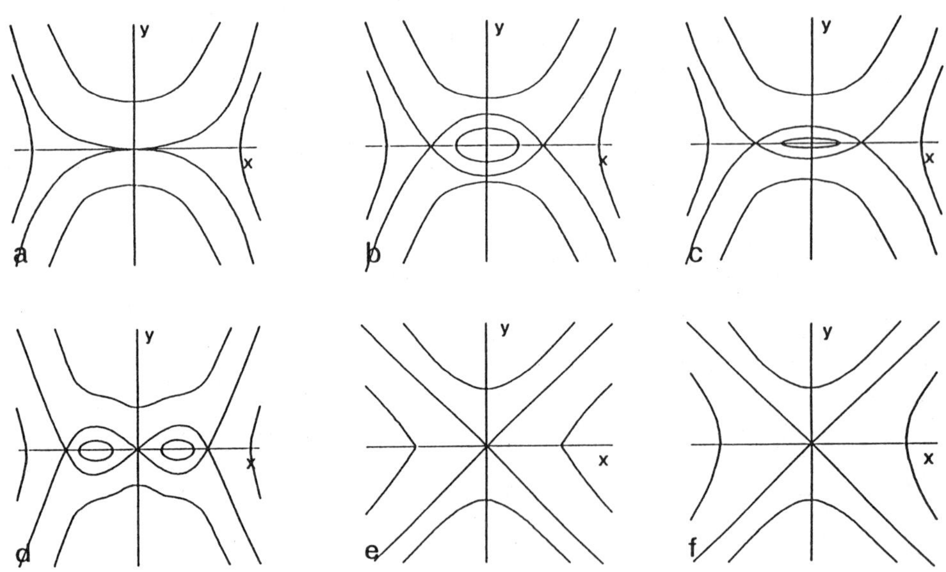

Figure 13 Period doubling case 2, $\alpha\beta < 0$.

Summarizing case 2, if $\nu < 0$ and $\alpha\beta < 0$, then as $\beta\mu$ increases through zero, we begin with a stable period-1 orbit with an unstable period-2 orbit nearby. The period-1 orbit goes unstable, sending out a stable period-2 orbit, which moves toward the unstable period-2 orbit; they collide and annihilate each other, leaving only the original period-1 orbit, which is now unstable. For $\alpha\beta < 0$, interchange "stable" and "unstable" in this description.

3 Computations

In order to illustrate the splitting of a pitchfork into a saddle-center bifurcation with a separate orbit close by, we generated Poincaré surfaces of section for the Hamiltonian H_1 near $e = -0.13$ with f small. As stated above, if $f = 0$ the potential energy has reflection symmetry, and a pitchfork bifurcation appears. For small but nonzero f, this symmetry is broken, so we might expect the nongeneric pitchfork bifurcation to be broken. The calculations told us that we have more to learn.

For $f = 0$, the SOS is shown in Figure 14a. We see a stable PO in the center with two unstable PO's symmetrically placed about it. Calculations show that as e increases, these two unstable PO's move toward the stable PO in the center, collide with it, and disappear, leaving the central orbit unstable. This is the standard "antipitchfork" bifurcation. Now we set $f = 0.001$. In (q, r) space the PO did indeed shift off the line $r = 0$. However the SOS retained its reflection symmetry about the p_r axis, and the antipitchfork structure was perfectly preserved for nonzero f (Figure 14b).

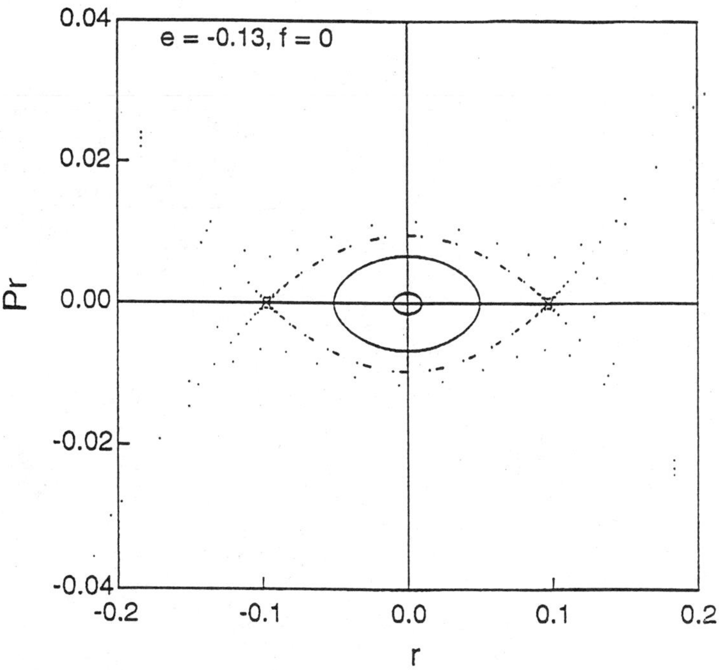

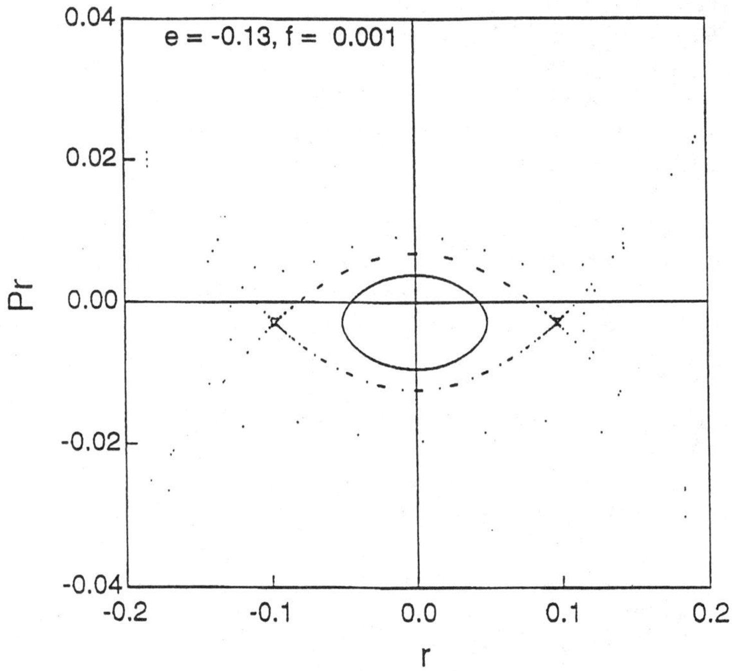

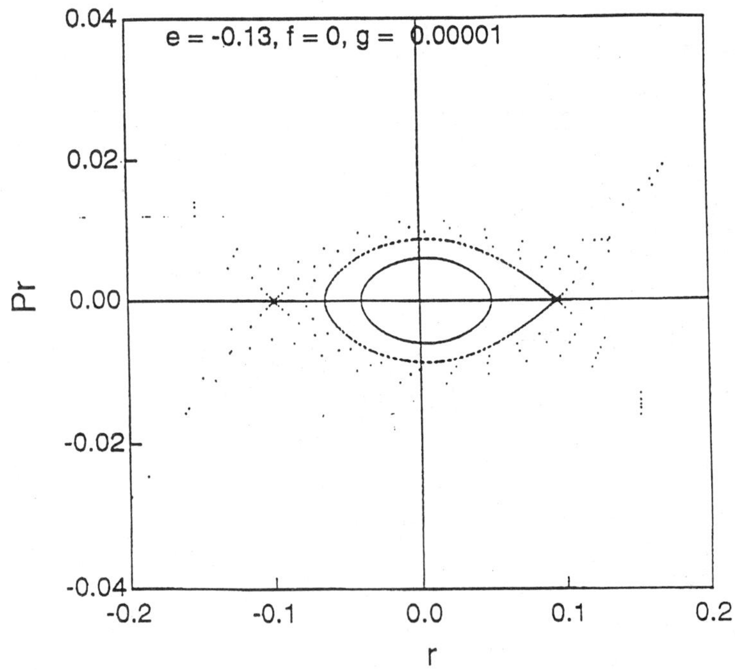

Figure 14 (a)(see Figure on previous page) Calculated Poincaré surface of section near an "antipitchfork" bifurcation. As e increases, the two x-points will move toward the o-point, collide with it and leave a single x-point in the center. (b) SOS for the Hamiltonian H_1 with small but nonzero f. The pitchfork structure is preserved. (c) SOS for the Hamiltonian H_2. The pitchfork is broken into one of the standard forms of a cusp bifurcation.

We might have anticipated this. In this case, the reflection symmetry ($r \to -r$) of the SOS is induced by the inversion symmetry of the potential energy, not the reflection symmetry. The parameter f breaks that reflection symmetry but not the inversion symmetry, so it preserves the pitchfork structure.

We decided to break all the symmetries by adding to the Hamiltonian another term,

$$H_2 = H_1 + 16g(q + r/2)^5.$$

(Such a term does not arise in atomic systems, but it illustrates the theory.) A numerically-determined SOS at $e = -0.13, f = 0, g = 1 \times 10^{-5}$ is shown in Figure 14c. It corresponds precisely to that predicted in Figure 4c.

A similar broken-pitchfork bifurcation was also studied by Prado and de Aguiar [1994], again by breaking symmetries in a two dimensional Hamiltonian. Another case was calculated by Schweizer et al. [1993]. They studied the bifurcations of the "parallel orbit" of an atomic electron in parallel electronic and magnetic fields. This is known to have an orderly sequence of pitchfork and period-doubling bifurcations. If the electric field is turned at small angle to the magnetic field, cylindrical symmetry is broken, the system becomes 3-dimensional, and the bifurcations have broken-pitchfork structure.

These calculations raise a fifth question, that should be added to the list in Section 1. (5) Can we list the types of two-parameter bifurcations that will be generic in the presence of symmetries? Specifically, we want a systematic way to go from symmetries of the Hamiltonian to symmetries of the normal form to symmetries of the SOS to symmetries of the bifurcations.

4 Conclusion

Experimental measurements of atomic absorption spectra display bifurcations of periodic orbits. These bifurcations often have the generic structures that are predicted by mathematical analysis. Some special, nongeneric bifurcations also occur because of the symmetries of the Hamiltonian. As symmetries are broken, nongeneric bifurcations turn into generic bifurcations in certain typical (generic) ways. Interpretation and prediction of these measurements will be aided by further development of the mathematical theory of bifurcations.

5 Acknowledgement

This research was supported by the Jeffress Foundation, the Office of Naval Research and the National Science Foundation, through grants to the College of William and Mary, and to the Institute for Theoretical Atomic and Molecular Physics at the Harvard-Smithsonian Center for Astrophysics. It is also supported by the Taft Foundation and a grant from the National Science Foundation to the University of Cincinnati.

Figure 2a was kindly provided to us by J. Main and G. Wiebusch, and Figure 2b by J. Shaw.

References

Abraham, R. and Marsden, J. [1978], *Foundations of Mechanics,* Benjamin-Cummings, London.

de Aguiar, M.A.M., Malta, C.P., Baranger, M., and Davies, K.T.R. [1987], Ann. Phys., **80**, 167.

Arnold, V.I. [1972], *Lectures on bifurcations in versal families,* Russian Math. Surv., **27**, 54–123.

Artin, M. [1968], *On the solutions of analytic equations,* Inventiones Math., **5**, 277–291.

Du, M.L. and Delos, J.B. [1988], Phys. Rev., **A38**, 1896, 1913.

Golubitsky, M. and Schaeffer, D. [1986], *Singularities and Groups in Bifurcation Theory I,* Springer-Verlag, New York.

Hale, J.K. and Chow, S-N. [1982], *Methods of Bifurcation Theory,* Springer-Verlag, New York.

Hilborn, R.C. [1990], *Chaos, An Introduction for Scientists and Engineers,* Oxford U. Press, New York.

Holle, A., Main, J., Wiebusch, G., Rottke, H., and Welge, K.H. [1988], Phys. Rev. Lett., **61**, 141.

Mao, J.-M. and Delos, J.B. [1992], Phys. Rev., **A45**, 1746.

Main, J., Wiebusch, G., Welge, K., Shaw, J. and Delos, J. B. [1994], Phys. Rev. A, accepted.

Meyer, K.R. [1970], *Generic bifurcation of periodic points,* Trans. Amer. Math. Soc., **149**, 95–107.

Meyer, K.R. [1971], *Generic stability properties of periodic points,* Trans. Amer. Math. Soc., **154**, 273–277.

Meyer, K.R. and Hall, G.R. [1992], *Introduction to Hamiltonian Dynamical Systems and the N-Body Problem,* Springer-Verlag, New York.

Moser, J.K. [1962], *On invariant curves of area-preserving mappings of an annulus,* Nachr. Akad. Wiss. Gottingen Math. Phys., **2**, 1-20.

Moser, J.K. [1968], *Lectures on Hamiltonian systems,* Mem. Amer. Math. Soc., **81**.

Moser, J.K. [1973], *Stable and Random Motion in Dynamical Systems,* Princeton University Press, Princeton, New Jersey.

Poincaré, H. [1899], *Les methods nouvelles de la mechanique celeste,* Gauthier-Villar, Paris.

Poston, T. and Stewart, I. [1978], *Catastrophe Theory and its Applications,* Pitman, Boston.

Prado, S. D. and de Aguiar, M.A.M. [1994], Ann. Phy. (NY), accepted.

Rimmer, R. [1983], *Generic bifurcation for involuntary area preserving maps,* Mem. Amer. Math. Soc., **41**, (272), 1–165.

Robinson, C. [1970], *Generic properties of conservative systems I & II,* Amer. Jour. Math., **92**, 562–603 & 897–906.

Robinson, C. [1971], *Generic one-parameter families of symplectic matrices,* Amer. Jour. Math., **93**, 116–122.

Scheizer, W., Niemeier, R., Wunner, G., and Ruder, H. [1993], Z. Phy., **D25**, 95.

Weinstein, A. [1972], *The invariance of Poincaré's generating function for canonical transformations,* Invent. Math., **16**, 202–213.

Fields Institute Communications
Volume 8, 1996

Homoclinic Orbits in the Forced Pendulum System

Daniel C. Offin and Hong-Fan Yu
Department of Mathematics
Queen's University
Kingston, Ontario, Canada K7L 3N6

1 Introduction

We study certain aspects of the forced mathematical pendulum whose governing equation is

$$\frac{d^2q}{dt^2} = -\sin(q) + f(t), q \in R, t \in R. \tag{1.1}$$

In particular we will be concerned with the existence of periodic and nonperiodic motions which are asymptotic to periodic motions. We place very mild restrictions on the function f assuming only that $f(\cdot)$ is continuous, T-periodic and of mean value zero on the period,

$$f(t+T) = f(t), \int_0^T f dt = 0$$

As a dynamical system the variables q, $\dot{q} = dq/dt$ and t are thought of as local variables in the extended phase space which is the quotient manifold $R/2\pi Z \times R \times R$, and our results will apply in this setting. The main theorem asserts that infinitely many homoclinic orbits to hyperbolic periodic orbits can be constructed as uniform limits of certain hyperbolic subharmonic orbits. Moreover, the homoclinic orbits are indexed by the winding number in configuration space $q \bmod 2\pi$. Our results apply to more general equations of the form

$$\ddot{q} + V_q(t,q) = f(t),$$

where $V(t+T,q) = V(t,q) = V(t,q+2\pi)$ and f satisfies the conditions stated earlier. However, for the sake of definiteness, we will use the forced pendulum equation throughout.

The mathematical pendulum or its variations has been the object of study for more than four centuries going back to Galileo in 1581. More recently there have been many authors who have studied various global properties of the flow

1991 *Mathematics Subject Classification.* Primary: 58F05, 70K40; Secondary: 34C25.
Research supported by an operating grant from NSERC of Canada.

Mawhin and Willem [1989], Fournier and Willem, Wiggins [1988], Offin [1990]. In particular we mention the apparently open question concerning the most general class of functions f for which equation (1.1) admits at least one T-periodic solution c.f. Mawhin [1982].

A common point of view for equation (1.1) is to consider $f(t)$ as a perturbation and indeed the dynamics of equation (1.1) as a nonintegrable perturbation of the integrable pendulum. This approach, which uses the perturbation technique due to Melnikov, applies when f is small say in the uniform norm. The conclusion of the method is that there is a transverse intersection of stable and unstable manifolds, resulting in horseshoes and infinitely many distinct homoclinic orbits embedded within the flow. A complete account of this approach may be found in Guckenheimer and Holmes [1983],Wiggins [1988].

By way of contrast with our results, we do not discuss the nonintegrability of equation (1.1) and indeed the character of transversality for the homoclinic orbits plays no role in the discussion. However, the main result we are reporting here coincides with the perturbation result, namely the existence of infinitely many homoclinic orbits. The novel point of our approach is that it is completely global and does not rely on a perturbation from the integrable case. Moreover, the homoclinic orbits have the special character of *multi-bump* homoclinic orbits investigated recently in the literature.

The technique which we employ here is variational with the periodic orbits corresponding to global minimizers of the action functional

$$A(q) = \int_0^T L(t, q, \dot{q})dt \ = \int_0^T 1/2\dot{q}^2 + \cos(q) + qf(t)\ dt$$

on the space of T-periodic functions H^1_T. The fact that $A(q)$ is bounded below on H^1_T provided that f has mean value zero is an observation of Willem (Mawhin and Willem [1989] pp14-16). In addition Mawhin and Willem show the existence of forced oscillations for pendulum like equations, where the potential is periodic in the q variables, in addition to mild growth conditions, and the forcing term has mean value zero on the period Mawhin and Willem [1989] (see Corollary 1.2). *This periodic orbit corresponds to a global minimizer of the action functional in the periodic function space* H^1_T. We shall use this observation in the sequel.

The variational argument for the existence of homoclinic orbits rests on the fact, proven in Offin [1990] , that the global minimizer of $A(\cdot)$, when nondegenerate, will correspond to a hyperbolic periodic orbit of equation (1.1). Nondegeneracy here refers to the nonexistence of periodic orbits for the linearized equations. The homoclinic orbits then are constructed as uniform limits of certain subharmonic orbits, also generically hyperbolic, which will be described below.

Rabinowitz [1990] [1994], Coti Zelati and Rabinowitz [1991] and Coti Zelati, Ekeland and Séré [1990] have also considered variational arguments for the existence of homoclinic orbits. In particular these authors obtain infinitely many homoclinic orbits in the case when the Hamiltonian is convex Coti Zelati, Ekeland and Séré [1990], or as in Coti Zelati and Rabinowitz [1991], the potenial is superquadratic. This condition however rules out the forced pendulum equation which has a periodic potential. On the other hand, Rabinowitz [1994] considers forced periodic potentials which are symmetric in the time variable. In this case he deduces that the global minimizers are symmetric as well as unstable. His result seems to be the first instance of global variational methods applied to the question of existence for homoclinic orbits to nontrivial periodic orbits without an explicit assumption of hyperbolicity as is the case, for example in Coti Zelati, Ekeland and Séré [1990].

2 Periodic Orbits and Their Properties

We will denote by $W^{1,2}(I,R)$ the function space of absolutely continuous functions with square integrable derivatives on the interval $I \subset R$. Denote $I_n = [-nT, nT]$ and the corresponding function space by E_n. Finally, we will denote by $H^1_{2nT} \subset E_n$, the space of periodic functions on the interval $I_n : x \in H^1_{2nT}$ if $x(-nT) = x(nT)$.

The equations of the forced pendulum, equation (1.1), are easily put into Hamiltonian form:

$$\dot{q} = \frac{\partial H}{\partial p} = p, -\dot{p} = \frac{\partial H}{\partial q} = \sin(q) - f(t) \tag{2.1}$$

with Hamiltonian $H(q,p,t) = \frac{1}{2}p^2 + (1-\cos(q)) + qf(t)$. To make H continuous we will work on the covering spacing space $R \times R \times R$, which we will refer to as the extended phase space for equation (2.1). In this case we let q_* denote the absolute minimizer, in the periodic function space H^1_T, of the functional $A(q)$:

$$A(q) = \int_0^T L(t,q,\dot{q})dt \ = \int_0^T 1/2\dot{q}^2 + \cos(q) + qf(t)\ dt,$$

$$A(q_*) = \inf_{H^1_T} A(q).$$

The existence of q_* of course depends on the function $f(\cdot)$ however as mentioned in the introduction, we specify at the outset certain conditions on f which guarantee existence of q_*. We will make an assumption concerning q_*, namely that q_* is *nondegenerate* and

$$A(q_*) = \inf_E A(q), E = W^{1,2}([0,T],R). \tag{2.2}$$

The point of this assumption is to remove the periodic boundary conditions which apply to the function space H^T_1. As mentioned in the introduction, the periodic function q_* is unstable. The assumption (2.2) is related to the global stable and unstable manifolds of q_*. This will be discussed further below.

The second variation of A, taken along q_*, is a quadratic functional on H^1_T. By the usual extension, we consider the symmetric bilinear functional on $H^1_T \times H^1_T$:

$$A''(q_*)(v,w) = \int_0^T \{\dot{v}\dot{w} - vw\cos(q_*)\}dt\ .$$

Since q_* minimizes A, then necessarily $A''(q_*) \geq 0$. We say that q_* is nondegenerate, as a minimizer of A, if $A''(q_*) > 0$. This condition is equivalent to the condition that q_* is nondegenerate as a periodic solution, namely, the variational equations along q_*:

$$\ddot{v} + \cos(q*)v = 0,$$

have no nontrivial T-periodic solutions. It is not difficult to see that a necessary condition for q_* to be nondegenerate is that

$$\int_0^T \cos(q_*)dt < 0,$$

simply take $v = w =$ constant. It is unknown however if this condition is sufficient for q_* to be nondegenerate. At any rate for the duration of this paper we assume that q_* is nondegenerate so that this condition remains in force.

We have the following characterization of nondegenerate minimizers c.f. Theorem 2.1 Offin [1990]

Theorem 2.1 *Nondegenerate minimizers of the functional $A(\cdot)$ on the space of periodic functions in H_T^1 correspond to hyperbolic periodic solutions of equation (2.1) on the extended phase space. In particular if P denotes the linearized Poincare mapping associated with the periodic orbit q_* then P has eigenvalues $e^{\lambda T}$ and $e^{-\lambda T}$ for some positive constant λ.*

According to Theorem 2.1, the periodic orbit q_* is unstable. In the extended phase space for equation (2.1) there are local stable and unstable invariant manifolds of orbits asymptotic to q_* in forwards and backwards time. It turns out that this fact is related to the assumption (2.2). As a consequence of Theorem 2.1 Offin [1990], it follows that the 2 dimensional invariant stable and unstable manifolds project diffeomorphically onto a tubular neighborhood of $(q_*(t), t)$ within $R \times R$. This projection property is related to and follows from a null rotation condition for these manifolds. In turn, the fact the the invariant manifolds have no points where the projection is singular is related to the vanishing of the Morse index, see Offin [1992] for more details. If this projection is global, that is, if the global stable and unstable manifolds project diffeomorphically onto $q(\bmod 2\pi) \times R$, our assumption (2.2) follows from standard arguments in the calculus of variations.

For future reference we mention one additional consequence of Theorem 2.1. Consider the second variation which as a bilinear form on $H_T^1 \times H_T^1$ will be denoted by J_0,

$$J_0(v, w) = \int_0^T \{\dot{v}\dot{w} - vw\cos(q_*)\}dt$$

The real parameter λ is an eigenvalue of J_0 if there is a nonzero T-periodic solution $v(t)$ of the periodic linear equation

$$\ddot{v} + (\cos(q_*) + \lambda)v = 0 \tag{2.3}$$

If v, λ is an eigenpair for J_0 then for every $w \in H_T^1$,

$$J_0(v, w) = \lambda \int_0^T vwdt$$

It is well known that there are infinitely many eigenvalues of J_0 tending to $+\infty$. The lowest eigenvalue, which will be denoted by λ_0, is nonnegative owing to the nonnegativity of J_0. Moreover the nondegeneracy assumption implies that λ_0 is strictly positive.

We consider eigenvalues of the functional J_n which by definition is the quadratic functional on the periodic function space H^1_{2nT} such that

$$J_n(v,w) = \int_{-nT}^{nT} \dot{v}\dot{w} - vw\cos(q_*)dt$$

In the following we continue to denote the lowest eigenvalue of J_n by λ_n, and the interval $[-nT, nT]$ by I_n.

Corollary 2.1 *The lowest eigenvalue of the functional J_n taken over the interval I_n is equal to λ_0, the lowest eigenvalue of J_0 taken over the interval [0, T]. Moreover the following inequality holds for $v \in H^1_T$*

$$J_n(v,v) \geq \lambda_0 \int_{-nT}^{nT} v^2 dt \tag{2.4}$$

Proof. We first observe that λ_0 is an eigenvalue of J_n. Indeed, if v_0 is a T-periodic solution of $\ddot{v} + (\cos(q_*) + \lambda_0)v = 0$, then a fortiori v_0 is a 2nT-periodic solution. Therefore λ_0 is an eigenvalue of J_n. It follows from the definition of λ_n that, $\lambda_n \leq \lambda_0$. We will show that equality holds in the circumstances of Theorem 2.1. To see this we observe that, since λ_0 is the lowest eigenvalue of J_0, the linear periodic equation (2.3) is hyperbolic for $\lambda < \lambda_0$. In particular, these equations have an exponential dichotomy for $\lambda < \lambda_0$ and hence no periodic solutions of period $2nT$. This means that there are no eigenpairs (λ_n ,v_n) for J_n with $\lambda_n <$ λ_0 . Consequently, $\lambda_n = \lambda_0$ for all $n \geq 1$. The final inequality is well known in the theory of quadratic functionals (see for example Hestenes [1966] pp.109-111).

Using this Corollary we obtain a fundamental inequality

Lemma 2.1 *There is a uniform positive constant β such that*

$$J_n(v,v) \geq \beta \int_{I_n} \dot{v}^2 + v^2 dt, v \in H^1_{2nT}$$

Proof. Since $J_n \geq 0$, suppose that, by way of contradiction, there is a sequence $v_m \in H^1_{2mT}$ such that

$$J_m(v_m, v_m) < 1/m \int_{I_m} \dot{v}_m^2 + v_m^2 dt.$$

Since J_m is quadratic on H^1_{2mT} , without loss of generality, we may assume that $\int_{I_m} \dot{v}_m^2 + v_m^2 dt = 1$. However from our earlier inequality (2.4), it follows that $\int_{I_m} v_m^2$ dt $\longrightarrow 0$, which in turn implies that $\int_{I_m} \dot{v}_m^2 dt \longrightarrow 1$. However, in this case

$$J_m(v_m, v_m) = \int_{I_m} \dot{v}_m^2 - (\cos q_*)v_m^2 dt \geq \int_{I_m} \dot{v}_m^2 - v_m^2 dt \longrightarrow 1$$

which gives a contradiction.

In the covering space of extended phase space, $R \times R \times R$, we have many other hyperbolic periodic solutions besides q_* namely

$$q_*(t) + 2k\pi, k \in Z.$$

However there is a much more interesting class of periodic solutions on the quotient space, namely, the subharmonic solutions having minimal period $2nT$, $n \in Z^+$. To construct these we will consider solutions in the covering space which satisfy the boundary conditions

$$q(nT) = q(-nT) + 2k\pi, \dot{q}(nT) = \dot{q}(-nT), k \in Z, (k,n) = 1 \tag{2.5}$$

The condition that k and n are relatively prime will ensure that the corresponding orbits on the quotient space have minimal period $2nT$. Because the boundary conditions (2.5) are not periodic in the covering space, we cannot immediately search for solutions using the variational technique with which we found q_*. This problem can be overcome however with a judicious choice of new coordinates.

In order to find solutions of equation (2.1) satisfying boundary conditions (2.5) we consider the following time dependent change of variables

$$q = q_*(t) + r_n(t) + x(t) \tag{2.6}$$

where by definition

$$r_n(t) = \int_{-nT}^{t} s(u)du,$$

and $s_n(t)$ satisfies the following conditions:

$S1)$ s_n is positive, even and periodic on $[-nT, nT]$

$S2)$ $s_n(t) = O(e^{\lambda(t-nT)^{-1}})$, as t $\longrightarrow nT$

$S3)$ $\int_{-nT}^{nT} s_n = 2k\pi$

Here we are fixing the integer k, and λ is the positive Floquet exponent related to the hyperbolic orbit q_*. From conditions S1)—S3), it follows that $r_n(-nT) = 0, r_n(nT) = 2k\pi$, and that $r_n(t) - k\pi$ is odd on the interval $[-nT, nT]$. Using the change of variables (2.6) and equations (2.1) we find that the induced equations in the variables $x, y = \dot{x}$ are

$$\dot{x} = y, \quad -\dot{y} = \sin(q_*(t) + r_n(t) + x(t)) - \sin(q_*) + \dot{s_n} \tag{2.7}$$

These equations are also Hamiltonian, with Hamiltonian K(x,y,t):

$$K(x, y, t) = 1/2y^2 - \cos(q_*(t) + r_n(t) + x(t)\ ) + x(\sin(q_*(t)) - \dot{s_n}\ )$$

Notice that K is periodic in t with (minimal) period $2nT$. We will look for $2nT$ periodic solutions of equations (2.7) as critical points in the periodic function class H^1_{2nT} of the functional $A_n(\cdot)$ which is defined as follows:

$$A_n(x) = \int_{-nT}^{nT} L(q_*(t) + r_n(t) + x(t)\ ) - L(q_*)dt = \text{constant}+$$

$$+ \int_{-nT}^{nT} 1/2\dot{x}^2 + \cos(q_*(t) + r_n(t) + x(t)\ ) - \cos(q_*) + x\{\sin(q_*(t)) - \dot{s_n}\ \}dt, \tag{2.8}$$

where the constant depends on r_n only and hence does not affect the critical points.

We remark that $\sin(q_*) - \dot{s}_n(t)$ has mean value zero on the period $2nT$. Indeed, since q_* is T periodic, integrating the equation (1.1) and using the fact that f has mean value zero on the period T, it follows that $\sin(q_*(t))$ has mean value zero. Moreover, $\dot{s_n}$ has mean value zero since s_n is 2nT- periodic. This fact together with spatial 2π-periodicity of the potential quarantees that the functional A_n has a critical point x_n which minimizes A_n over the function class H^1_{2nT} (see the discussion of this result of Mawhin, Willem in the introduction). Notice that if k and n are relatively prime, then s_n and x_n have minimal period $2nT$.

Finally if $q_n = q_*(t) + r_n(t) + x(t)$, then it is an easy calculation to show that the trajectory q_n together with its derivative $\dot{q}_n$ satisfies the boundary conditions (2.5). Thus we have proven the following theorem.

Theorem 2.2 *For every integer n which is relatively prime to a given integer k, there exists a periodic solution q_n, p_n of minimal period $2nT$ on the quotient manifold $R/2\pi Z \times R \times R$ for the differential equations (2.1). Moreover these periodic solutions are hyperbolic whenever they are nondegenerate, that is, whenever the corresponding variational equations have no nontrivial $2nT$ periodic solutions.*

We now wish to establish certain convergence properties of the periodic orbits q_n, as $n \to \infty$, whose existence is asserted in Theorem 2.2. The limit of these orbits can be considered when they stay in some bounded neighborhood of the unstable orbit q_*. The following result describes this situation, whereby the orbits q_n in the covering space $R \times R$ do not wander too much from the orbit q_*. In the following discussion, we will denote by q_n the corresponding orbits on the covering space $R \times R \times R$.

Proposition 2.1 *The function $q_n(t)$ intersects the functions $q_*(t)$ and $q_*(t) + 2k\ \pi$ exactly once. This intersection with $q_*(t)$ will occur at time t_n and with $q_*(t)\ + 2k\ \pi$ at time $t_n\ +\ 2nT$, where $-3nT < t_n \leq -nT$. Moreover, if x_n is related through r_n and equation (2.6) to q_n, then*

$$\mid x_n(t) \mid \leq 4k\pi$$

Proof. The fact that q_n must intersect q_* and $q_* + 2k\pi$ follows by periodicity of q_* and the fact that $q_n = q_*(t) + r_n(t) + x(t)$, with x_n periodic and $r_n \longrightarrow \pm\infty$, as $t \longrightarrow \pm\infty$. The crossings of q_* and $q_n + 2k\pi$ can occur only once comes from the fact that both q_* and q_n are hyperbolic periodic solutions and hence must satisfy a certain minimality condition with respect to the action functional A. Namely, if $P(t_1, q_*(t_1))$ and $Q(t_2, q_*(t_2)$ denote two fixed points on graph q_*, consider the functional A:

$$A(q) = \int_{t_1}^{t_2} 1/2\dot{q}^2 + \cos(q) + qf(t)dt$$

taken over all piecewise smooth arcs joining P and Q. Since q_* is hyperbolic with real Floquet multipliers, then $q_* \mid_{[t_1,t_2]}$ is a strong local minimizer of A(q). The same minimality principle holds for the trajectories q_n. It is this minimality property which prevents additional crossings.

Indeed, if q_n intersected q_* more than once, the minimality principle stated above would be violated since minimizers of the action functional must be smooth curves. Moreover, multiple intersection points would allow the minimizing curve to be taken along either q_* or q_n between their points of intersection. Since their tangent vectors at a point of intersection do not agree, this would result in a nonsmooth minimizer for the action. Therefore we conclude that the curves q_* and q_n intersect only once.

Suppose q_n and q_* intersect at t_1, then q_n intersects $q_* + 2k\pi$ at $t_1 + 2nT$. Let $t_1 = t_0 + 2lnT$, with l an integer, and $t_0 \in [-3nT, -nT]$. For any $t \in [t_1, t_1 + 2nT]$, we have $t_* \in [t_0, t_0 + 2nT]$, with $t_* = t - 2lnT$. Moreover,

$$\begin{aligned} q_n(t) &= q_*(t_* + 2lnT) + x_n(t_* + 2lnT) + r_n(t_* + 2lnT) \\ &= q_*(t_*) + x_n(t_*) + r_n(t_*) + 2lk\pi \end{aligned}$$

Because of $2k\pi \geq r_n(t) \geq -2k\pi$, with $t \in [t_0, t_0 + 2nT]$, we have;

$$\mid x_n(t) + 2lk\pi \mid \leq 4k\pi.$$

With l a fixed integer, the solution of (2.7)

$$x_n(t) + 2lk\pi$$

will replace $x_n(t)$ in the change of variables (2.6).

Our goal is to show that the periodic solutions q_n converge, as $n \to \infty$, to a homoclinic solution having winding number (mod 2π) equal to k. In order to consider the limiting properties of q_n, as $n \longrightarrow \infty$, we must first place some restrictions on the curves r_n which were used to construct q_n. We want these curves to converge to a (nonperiodic) curve on R which has the rotation number (mod 2π) equal to k. We define the following nonperiodic curve r(t) on the real line R, and which satisfies the following conditions: $r(t) = \int_{-\infty}^{t} s(u)du$,

$$\begin{aligned} &S1') \quad s \in L^{\alpha}(R), \text{ for each } \alpha \geq 1, \text{s is positive and even on R} \\ &S2') \quad s(t) = O(e^{-\lambda t}), \text{ as } t \longrightarrow \infty \\ &S3') \quad \textstyle\int_R s(u)du = 2k\pi \end{aligned}$$

Here λ is the positive Floquet exponent for the hyperbolic periodic orbit q_*. Observe that $r(t) - k\pi$ is an odd function on R. The connection with the sequence r_n will be in terms of convergence in $H^1(R)$. Specifically, we choose the r_n so that they converge to $r(t)$ in $H^1(-\infty, 0] \cap L^{\alpha}(-\infty, 0]$, for all $\alpha \geq 1$. That is we will assume that

$$\int_{-\infty}^{0} (\tilde{r}_n - r)^{\alpha} + (\tilde{s}_n - s)^{\alpha} dt \longrightarrow 0, \text{ as } n \longrightarrow \infty$$

where

$$\tilde{r}_n = \begin{cases} 0, & \text{if } -\infty < t \leq -nT \\ r_n(t), & if{-nT} \leq t \leq 0 \end{cases}$$

with a similar description for $\tilde{s}_n$. Based on these considerations we will state a simple estimate which will be used in the considerations of convergence for the periodic orbits q_n.

Lemma 2.2 *There is a uniform constant $M < \infty$ such that*

$$\int_{-nT}^{nT} \{\cos(r_n + q_*) - \cos(q_*)\}dt < M$$

$$\int_{-nT}^{nT} \{\sin(r_n + q_*) - \sin(q_*)\}dt < M$$

Proof: Let $\int_{-nT}^{0}\{\cos(r_n + q_*) - \cos(q_*)\}dt \leq M/2$, then

$$\int_{-nT}^{0} \{\cos(r_n + q_*) - \cos(q_*)\}dt \leq \int_{-nT}^{0} r_n dt \leq M/2$$

and

$$\int_{0}^{nT} \{\cos(r_n + q_*) - \cos(q_*)\}dt \leq \int_{0}^{nT} (2k\pi - r_n)dt \leq M/2.$$

In order to show convergence of the q_n it will be necessary to obtain certain estimates on the critical values $A_n(x_n)$. As a first step in this direction, we will show that these critical values are bounded independently of n. To make this more transparent we will introduce a new functional which is essentially the translate of the action functional centered at the global minimizer q_*. Denote $q = q_* + \xi$, and define the functional B(ξ),

$$B(\xi) = \int_{0}^{T} \{1/2\dot{\xi}^2 + \cos(\xi + q_*) - \cos(q_*) \ + \xi\sin(q_*)\}dt \tag{2.9}$$

The same integrand taken over the interval $I_n = [-nT, nT]$ will be denoted by $B_n(\xi)$, where $\xi \in H^1_{2nT}$.

Lemma 2.3 *$B(\xi) > 0$ for all $\xi \in H^1_T$ unless $\xi = 2l\pi$. Similarly, $B_n(\xi) > 0$ for all $\xi \neq 2l\pi$ which belong to H^1_{2nT}.*

Proof: We compute the translate of the action functional A

$$\begin{aligned} A(\xi + q_*) - A(q_*) &= \textstyle\int_0^T (1/2\dot{\xi}^2 + \dot{\xi}\dot{q}_* + \cos(\xi + q_*) - \cos(q_*) + \xi f = dt \\ &= \textstyle\int_0^T (1/2\dot{\xi}^2 + \cos(\xi + q_*) - \cos(q_*) + \xi\sin(q_*)) = dt \\ &= B(\xi) \end{aligned}$$

Since q_* is a global minimizer on H^1_T, $B(\xi) > 0$, unless $\xi = 2l\pi$. Moreover since q_* is a nondegenerate minimizer of A taken over the interval $[0, T]$, the $2n$-th cover of q_* taken over the interval I_n will be a nondegerate minimizer of A taken over the interval I_n (see for example Lemma 3.1 Offin [1990]). This property implies the final inequality of the Lemma.

□

Proposition 2.2 *There are uniform constants m, M such that the critical values c_n of the functional A_n, $c_n = A_n(x_n)$ satisfy the estimates*

$$m \leq c_n \leq M$$

Proof: First, since x_n is the global minimizer in H^1_{2nT} of the functional A_n,

$$A(x_n) \leq A_n(0) = \int_{I_n} \{\cos(r_n + q_*) - \cos(q_*)\}dt \leq M,$$

by virtue of Lemma 2.2. Moreover,

$$A_n(x_n) = B_n(x_n) + \int_{I_n} \{\cos(x_n + r_n + q_*) - \cos(x_n + q_*) - x_n\dot{s}_n\}dt. \quad (2.10)$$

Now $B_n(x_n) > 0$ by the lemma 2.3, and the remaining integral is bounded below by the integral

$$-\int_{I_n} \min\{r_n, 2k\pi - r_n\} + | x_n || \dot{s}_n | \, dt.$$

This expression in turn is uniformly bounded below, by m say, due to the fact that $| x_n | \leq 4k\pi$ (see Proposition 2.1).

□

We are now in a position to describe the limiting properties of the periodic trajectories x_n, and by way of the change of variables (2.6), the corresponding solutions q_n of the differential equation (1.1). The limiting properties of x_n are intimitely tied to the limiting properties of the critical values c_n and thus to the limit of the action functionals A_n, as $n \longrightarrow \infty$. We will first describe this limiting action functional which will play a significant role in the proof of our main theorem.

Consider the functional A_∞ defined on $H^1_{loc}(R)$

$$A_\infty(x) = \lim_{n\to\infty} \int_{I_n} L(x + r + q_*) - L(q_*)dt =$$

$$= \lim_{n\to\infty} \int_{I_n} 1/2\dot{x}^2 - V(t, x)dt + constant,$$

where

$$V(t, x) = -\cos(x + r + q_*) + \cos(q_*) - x\{\sin(q_*) - \dot{s}\},$$

and r, s are defined by conditions $S1'), \cdots, S3')$. We remark here that the limit in the defintion of A_∞ exists (possibly $+\infty$) because the sequence $A_n(x)$ is positive and increasing, due to assumption (2.2).

It is not difficult to see that $\| x_n \|_2$ is bounded in $H^1_{loc}(R)$ and hence that some subsequence of x_n converges weakly in $H^1_{loc}(R)$. Indeed,

$$A_n(x_n) = 1/2 \| \dot{x} \|^2_{L^2(I)} + \text{constant}$$

where the constant depends only on the length of the interval $I = I_n$, since $\| x_n \|_\infty$ is uniformly bounded by Proposition 2.1. Since $A_n(x_n)$ is bounded (Proposition 2.2), it follows that $\| x_n \|_{H^1(K)}$ is bounded for any bounded interval K. We will denote the limiting trajectory by $x_\infty \in H^1_{loc}(R)$.

Lemma 2.4 *$A_\infty(x_\infty)$ is finite.*

Proof. The functional A_k is weakly lower semi-continuous so that

$$A_k(x_\infty) \leq \lim_{n\to\infty} A_k(x_n) \leq M$$

Therefore,

$$\lim_{k\to\infty} A_k(x_\infty)$$

is bounded.

Lemma 2.5 *If $n > m$ are positive integers, then, $B(x_\infty) \mid_{I_n - I_m} \to 0$, as $n,\ m \to 0$*

Proof. Since

$$A_n(x) = A_n(x) \mid_{I_n-I_m} + A_n(x) \mid_{I_m}$$

and both summands are positive, $A_n(x_\infty) \mid_{I_n-I_m} \to 0$, as m,n$\to \infty$, follows from the fact that $\lim_{n\to\infty} A_n(x_\infty)$ is finite.

Let us define a functional $\delta_{n,m}(x)$ according to the rule

$$\delta_{n,m}(x) = A_n(x) \mid_{I_n-I_m} - B(x) \mid_{I_n-I_m}$$

Then it is easy to see, based on the properties $S1'), \cdots, S3')$ of the function r, that $\delta_{n,m}(x) \to 0$ uniformly for $x \in H^1_{2nT}$. Since

$$A_n(x_n) \longrightarrow 0$$

on $I_n - I_m$ it follows that

$$B(x_\infty) \mid_{I_n-I_m} \longrightarrow 0$$

as n,m$\to \infty$.

We want to show how this implies that the H^1_{2nT} norms of x_∞ are bounded. For this, we will introduce an equivalent norm on H^1_{2nT}, denoted $\| x \|_B$, and defined by

$$\| x \|_B^2 = \int_{I_n} \dot{x}^2 - x^2 \cos(q_*)\, dt$$

The quadratic functional on the right hand side is the second variation of the functional B, centered at 0. The norm equivalence of this expression with the usual norm on H^1_{2nT} is guaranteed by Lemma 2.1.

We have the following relation between the functional B and its second variation:

$$B(x) = 1/2 \| x \|_B^2 + \int_{I_n} \cos(x+q_*) - \cos(q_*) + x\sin(q_*) + 1/2x^2\cos(q_*)dt. \tag{2.11}$$

$$B(x) = 1/2 \| x \|_B^2 - \int_{I_n} x^3 \sin(q_* + u)dt,$$

where $0 \leq |u| \leq |x(t)|$. From this last relation we may deduce the following statement:

Lemma 2.6 $B(x) = 1/2 \parallel x \parallel_B^2 + o(\parallel x \parallel_B^2)$, *as* $x \to 0$ *in* H^1_{2nT}

Of course, the statement of Lemma 2.6 also holds on the interval $I_n - I_m$ and consequently, we can now prove:

Proposition 2.3

$$\parallel x_\infty \parallel_B^2 |_{I_n}$$

are bounded independent of n.

Proof. As noted above, $B(x_\infty) \mid_{I_n - I_m} \to 0$, as n,m$\to \infty$. Since $B(x) > 0$, Lemma 2.6 implies that $\parallel x_\infty \parallel_B^2 \to 0$ on the intervals $I_n - I_m$.

The limiting trajectory x_∞ is a critical point of the functional A_∞ which corresponds to a minimum value of A_∞. This follows from the fact that x_n are global minimizers of A_n and that $A_\infty = \lim A_n$. We are being a little imprecise here, because x_n are periodic functions and therefore $A_\infty(x_n) = +\infty$. To make this argument correct, we have to modify x_n outside the interval I_n so that the contribution to A_∞ on larger intervals is negligible. This is possible to do, for example, define the extended trajectories $\tilde{x}_n(t)$

$$\tilde{x}_n(t) = \begin{cases} x_n(t) & t \in I_n \\ 0 & t \notin I_n \end{cases}$$

Notice that if $\tilde{x}_n(t)$ is continuous, then, it belongs to $H^1(R) \bigcap H^1_{2nT}$. In case $x_n(\pm nT) = 0$, then $\tilde{x}_n$ is continuous on R. Otherwise, $x_n(\pm nT) = a_n \to 0$ as $n \to \infty$. In this case we define $\tilde{x}_n(t)$ differently so that $\tilde{x}_n(t) = 0$ for $\mid t \mid > (n+1)T$, and $\tilde{x}_n(t)$ is continuous on R.

Proposition 2.4 A_n *is bounded below on* $H^1(R)$, $\tilde{x}_n$ *is a minimizing sequence, and*

$$A_\infty(x_\infty) = \inf A_\infty$$

Proof. Let $x \in H^1(R)$, then $A_\infty(x) \geq 0$ by virtue of the assumption (2.2).

To see that $\tilde{x}_n(t)$ is a minimizing sequence, fix $\varepsilon > 0$, and choose $x^* \in H^1(R)$ so that

$$\inf A_\infty < A_\infty(x^*) < \inf A_\infty + \varepsilon.$$

We can modify x^* by choosing n so large that

$$A_\infty(x^*) \mid_{R \setminus I_n} < \varepsilon.$$

Now define x_ϵ on R so that for some $\alpha(\varepsilon) < T$,

$$x_\epsilon(t) = \begin{cases} x^* & \mid t \mid < (n-1)T \\ 0 & \mid t \mid > (n-1)T + \alpha(\varepsilon) \text{ and} \end{cases}$$

$$A_\infty(x_\epsilon) < A_\infty(x^*) + 3\varepsilon.$$

Then $x_\epsilon \in H^1_{2nT}$ so

$$A_\infty(x_\infty) \le \lim_{n\to\infty} A_n(x_n) < \lim_{n\to\infty} A_n(x_\epsilon)$$
$$= A_\infty(x^*) + 3\varepsilon < \inf A_\infty + 4\varepsilon$$

Since $\epsilon > 0$ is arbitrary, it follows that $\tilde{x}_n(t)$ is a minimizing sequence for A_∞ on $H^1(R)$, and that x_∞ minimizes A_∞ on $H^1(R)$.

The relation of the critical points of A_∞ to solutions of the pendulum equation (1.1) is contained in the following Lemma.

Lemma 2.7 *If $x \in H^1(R)$ is a critical point of the functional A_∞, then $q = x + r + q_*$ is a classical solution of equation (1.1) such that*

$$| q(t) - q_*(t) | \to 0, | \dot{q}(t) - \dot{q}_*(t) | \to 0$$

as $t \to \pm\infty$.

Our main result concerns the convergence of the trajectories q_n to a nontrivial homoclinic orbit. We introduce the notation

$$q_\infty(t) = x_\infty(t) + r(t) + q_*(t)$$

for the limiting trajectory in configuration space. Recall that q_∞ depends implicitly on the integer k which describes the winding number of the trajectory q_∞.

Theorem 2.3 *Assume that condition (2.2) holds and that the periodic orbit q_* is nondegenerate. On the quotient manifold $R/2\pi Z \times R \times R$, there exist infinitely many distinct homoclinic orbits asymptotic to the hyperbolic T-periodic orbit $(q_*, \dot{q}_*, t)$. These orbits are indexed by an integer k which is the winding number in the configuration space $q(\text{mod } 2\pi)$.*

Proof. On the basis of Proposition 2.4 and Lemma 2.7, we may conclude that q_∞ corresponds to a nontrivial homoclinic orbit with winding number k. The nontriviality of q_∞ comes from the fact that q_n are homotopically nontrivial on the configuration space q(mod 2π). Finally we want to show that we get a distinct homoclinic orbit for each choice of the integer k.

The problem which must be overcome is that in case $k' = mk$, with $m > 0$, it might be possible that the corresponding homoclinic orbit q'_∞ with winding number k' is a m-multiple cover of a homoclinic orbit with winding number k.

This phenomenon occurs in the unforced pendulum equation where the rotational periodic orbits (in the covering space) with winding number k converge uniformly to k-multiple covers of the fundamental separatrix solution, together with $k-1$ copies of the unstable critical point $x = \pi$. In this case that is, the periodic rotational orbits converge to a disjoint union of orbits which include homoclinic orbits of smaller rotation number. In the case of the unforced pendulum, however, the rotational solutions do not have the minimal periodicity property which the periodic orbits q_n have been shown to enjoy. Indeed, for the unforced pendulum, there are rotational solutions of winding number k having arbitrarily small period. This situation does not arise in the forced problem since the trajectories q_n have minimal period $2n\pi$.

To rule out the case that q_∞ is a multiple covering of homoclinics with smaller rotation number, we observe that if this were true, the orbit of q_∞ should correspond to a multiple union of disjoint orbits of solutions for the equations (2.1).

However, it is shown in Proposition 2.4 and Lemma 2.7 that q_∞ is a classical solution of equation (1.1). In particular, q_∞ cannot have an orbit which corresponds to different solutions of (2.1). As explained above, this rules out the possibility of multiple covers, and completes the proof that there is a nontrivial homoclinic orbit for each rotation number k.

References

Coti Zelati, V., Ekeland, I. and Séré, E., [1990], *A variational approach to homoclinic orbits in hamiltonian systems*, Math. Annalen, **288**, 133-160.

Coti Zelati, V. and Rabinowitz, P.H. [1991], *Homoclinic orbits for second order hamiltonian systems possessing superquadratic potentials*, J. Amer. Math. Soc., **4**, 693-727.

Fournier, G. and Willem, M., *Multiple solutions of the forced double pendulum equation*, Ann. de l'Institute H. Poincaré, Analyse Non Linéaire, to appear.

Guckenheimer, J. and Holmes, P. [1983], *Nonlinear oscillations, dynamical systems and bifurcations of vector fields*, Springer Verlag, New York.

Hestenes, M.R. [1966], *Optimal Control and the Calculus of Variations*, Wiley-Interscience, New York.

Mawhin, J. [1982], *Periodic oscillations of forced pendulum-like equations*, in Ordinary and Partial Differential equations, Conference proceedings, Dundee 1982; (eds. Everitt and Sleeman), Lecture Notes in Math. 964 Springer Verlag, Berlin, 458-476.

Mawhin, J. and Willem, M. [1989], *Critical point theory and hamiltonian systems*, Springer Verlag, New York.

Offin, D.C. [1990], *Subharmonic oscillations for forced pendulum type equations*, Differential and Integral Equations, **3**, 965-972.

Offin, D.C. [1992], *A spectral theorem for reversible second order equations with periodic coefficients*, Journal of Differential and Integral Equations, **5**, 615-629.

Rabinowitz, P.H. [1990], *Homoclinic orbits for a class of hamiltonian systems*, Proceedings Roy. Soc. Edinburgh, **114A**, 33-38.

Rabinowitz, P.H. [1994], *Heteroclinics for a reversible hamiltonian system*, Ergod. Th. and Dynamical Sys., **14**, 817-829.

Wiggins, S. [1988], *Global bifurcations and chaos*, Springer Verlag, New York.

Fields Institute Communications
Volume 8, 1996

Dynamics Near Stable Relative Equilibria at Non-Generic Momenta: a Numerical Investigation

George W. Patrick
Department of Mathematics and Statistics
University of Saskatchewan
Saskatoon, Saskatchewan
Canada, S7E 5E6

1 Introduction

Suppose that one is given a Hamiltonian system with symmetry (P, ω, G, H); that is, suppose that (P, ω) is a symplectic manifold on which a compact Lie group G acts symplectically, let $H : P \to \mathbb{R}$ be a G-invariant Hamiltonian function, and let $J : P \to \mathfrak{g}^\star$ be an $\mathrm{Ad}^\star$ equivariant momentum map. See Abraham and Marsden [1978] for the basic definitions. A relative equilibrium is an element of phase space $p_e \in P$ with Hamiltonian evolution coincident with the action of a one parameter orbit of the symmetry group G, so that the evolution of p_e is

$$t \mapsto \exp(\xi_e t) \cdot p_e,$$

for some $\xi_e \in \mathfrak{g}$. For example, the Lagrange equilateral triangle solutions for the planar 3 body problem are relative equilibria, as are circular motions of a particle in a central potential. Relative equilibria, being explicit nontrivial solutions of the evolution equations, are of obvious interest when examining the dynamics of a Hamiltonian system with symmetry.

A relative equilibrium p_e is regular if p_e admits no infinitesimal symmetries; that is, if for all $\xi \in \mathfrak{g}$,

$$\xi(p_e) = \xi \cdot p_e = \left.\frac{d}{dt}\right|_{t=0} \exp(t\xi) \cdot p_e \neq 0.$$

Regular relative equilibria correspond to fixed points of the flow on the Poisson or symplectic reduced phase spaces. A regular relative equilibria is formally stable if the Hessian of the reduced Hamiltonian at the corresponding point of the reduced system is definite, or equivalently, if the Hessian of $H - J_{\xi_e}$ is definite on a subspace of $\ker dJ(p_e)$ that is complementary to the group directions $T_{p_e}(G_{\mu_e} \cdot p_e)$; see, for

1991 *Mathematics Subject Classification.* Primary: 70E15, 70H33, 70K20; Secondary: 58F05, 65L07, 70-04.

Research supported by NSERC operating grant OGP0105716.

example Holm, Marsden, Ratiu and Weinstein [1984], Marsden, Simo, Lewis and Posbergh [1989], and Simo, Lewis and Marsden [1991]. If $\tilde{G}$ is a subgroup of G, a relative equilibria p_e is stable modulo $\tilde{G}$ if any evolution beginning sufficiently close to p_e remains arbitrarily close to $\tilde{G} \cdot p_e$, for all time. In Patrick [1992], one finds the following theorem:

Theorem 1 *A formally stable relative equilibria with momentum $J(p_e) = \mu_e$ is stable modulo G_{μ_e}, where G_{μ_e} is the isotropy group of the co-adjoint action of G on $\mathfrak{g}^\star$.*

Thus, given a particular regular relative equilibria, the presence or absence of formal stability has a well defined dynamic interpretation.

To interpret theorem (1), note that, by conservation of momentum and $\mathrm{Ad}^\star$ equivariance,

$$\exp(\xi_e t) \cdot \mu_e = \mu_e,$$

so that the one parameter group generated by ξ_e is contained in G_{μ_e}. Now, generically, G_{μ_e} is a maximal torus of G and the one parameter group generated by ξ_e is a maximal torus, so that, generically, the closed subgroup generated by ξ_e is a maximal torus and is coincident with G_{μ_e}. Thus, motions of the system sufficiently near a generic, formally stable relative equilibrium remain arbitrarily near the closure of the orbit $\{ \exp(\xi_e t) \cdot p_e \mid t \in \mathbb{R} \}$, a situation that corresponds to the orbital stability of Birkoff (see, for example Abraham and Marsden [1978]). Generically, then, that a relative equilibrium is merely stable modulo G_{μ_e} is not easily discernible: an observer viewing a perturbation of the relative equilibrium over a moderate time will mostly perceive the closure of $G_{\mu_e} \cdot p_e$, whereas that set is fixed in time. If one regards the deviation at time t of a perturbation of p_e from the point $\exp(\xi_e t) \cdot p_e$ as *drift*, then the drift of a generic, formally stable relative equilibrium could be characterized as *trivial*, or perhaps more vividly, as *time-like*.

Consider, for example, the free motion of a single rigid body. There, the relative equilibria are the rotations of the body about it's principle axes of inertia, and the only relative equilibria with nongeneric momenta are the points of phase space corresponding to arbitrary fixed configurations. Since the symplectic reduced space associated to zero total angular momentum is a point, those nongeneric relative equilibria are formally stable. Yet, a small perturbation of one such fixed configuration will cause a slow movement that can be arranged to pass close to configurations not all contained in any proper subgroup of the symmetry group $SO(3)$, so theorem (1) is the best possible.

Thus, if one must adhere to an edict of Birkoff orbital stability, then formally stable relative equilibria at non-generic momentum may be *unstable*. Of course, there are degrees and forms of instability. A most severe kind of instability occurs at relative equilibria corresponding to linearly unstable equilibria on the reduced space. An arbitrarily small perturbation of such a relative equilibrium will leave some neighborhood of the group orbit $G \cdot p_e$; the motion thereafter, at least for some time interval, may not appear as any relative equilibrium would, since the perturbed system may evolve to points in phase space that are not near to relative equilibria. In contrast, for formally stable relative equilibria at non-generic momenta, theorem (1) implies a less severe kind of instability: the system does not remain near to the orbit of the relative equilibrium p_e, but evolves to phase space points near the set of relative equilibria $G_{\mu_e} \cdot p_e$. An observer viewing the evolution of a perturbation of such a relative equilibrium over a moderate time perceives not only the closure of the orbit $\{\exp(\xi_e t) \cdot p_e \mid t \in \mathbb{R}\}$, given that the perturbation is small, but also perceives a slow change in that set in a way that conforms with the attendant symmetry of the system. This situation could be characterized by the term *nontrivial drift.*

This article describes some numerical experiments concerning drift near relative equilibria with non-generic momenta, and for which $\xi_e \neq 0$. The purpose of this work is to expose nontrivial drift as a real phenomena and to delineate the way that the drift depends on initial perturbations. This work is towards to a theoretical analysis of the drift phenomena, which will be published elsewhere in Patrick [1995].

Specifically, we consider here the mechanical system two free axially symmetric rigid bodies coupled by an ideal ball and socket joint. This system will be denoted by the acronym 2ACRB. In a state of relative equilibrium, 2ACRB evolves such that the angle between the two bodies at the joint is fixed, the entire system rotates at some constant angular velocity, and the bodies themselves spin at constant (but perhaps different) rates. Because of the presence of multiple rotating parts, this system has formally stable relative equilibria which are not equilibria, and which have zero total angular momentum. Section (2) precisely defines this system and the relevant relative equilibria. In viewing animations of 2ACRB, nontrivial drift is readily apparent, and easily described. For small perturbations of a relative equilibrium, and over moderate time, so that the system undergoes moderately many rotations, one observes nothing especially noteworthy: the motion appears to be a quasiperiodic perturbation of the relative equilibrium, a circumstance that is expected, since such a perturbation will excite near linear modes at the corresponding stable relative equilibrium in the reduced space. However, if the perturbed system has nonzero total angular momentum and is observed for a long time, the entire system is seen to slowly rotate along the axis of the angular momentum of the perturbed system. Interestingly, this slow motion can be retrograde: for the parameter values and relative equilibrium discussed in this article, the sense of the slow rotation is opposite to the total angular momentum of the perturbed system.

To investigate the drift phenomena requires quantifying it. Abstractly, drift can be quantified as follows: Suppose that $p_i \in P$, $i \geq 0$ is approximately the state of the system at time $i\Delta t$, given that the system is following an evolution from a perturbation of a formally stable relative equilibrium p_e. By theorem (1), the system will evolve near the group orbit $G_{\mu_e} \cdot p_e$, so that the p_i may be approximated by $g_i p_0$, $g_i \in G_{\mu_e}$. Over moderate time, the evolution will be a perturbation of a relative equilibrium, so that over subsequences of moderate length the group elements g_i will be near to some right translation of a one parameter subgroup of G_{μ_e}. Finding those one parameter subgroups for successive such subsequences yields a sequence of elements in the Lie algebra $\mathfrak{g}$ that reflects the drift, since it is the discrete evolution corresponding to the adjoint action of the drift on the element ξ_e. In the case of 2ACRB, this evolution in the Lie algebra turns out to be that of rotation about a single vector, called the drift vector. The details of these calculations for 2ACRB are explained in section (3). The dependence of drift on the initial perturbation is investigated in section (4). The main result is that the drift vector depends only on the difference of the perturbed and unperturbed momenta, and depends nearly linearly and is parallel or opposite to that difference.

Given the presence of formal stability, one may wonder at the origins of the instability described above, and section (4) closes with a discussion of what may be involved. Let p_e be a relative equilibrium with Lie algebra element ξ_e. Consider the vector field $X_{H_{\xi_e}}$ that is the difference of the Hamiltonian vector field and the infinitesimal generator corresponding to ξ_e. By definition $X_{H_{\xi_e}}(p_e) = 0$, so that $X_{H_{\xi_e}}$ admits a linearization (Abraham and Marsden [1978]) at p_e. For the specific relative equilibrium of 2ACRB considered in this article, that linearization has purely complex eigenvalues and a nontrivial nilpotent part. The nilpotent part of the linearization can be seen to descend to a symmetric linear map from $\mathfrak{g}^\star$ to $\mathfrak{g}$ and an eigenvalue of that map closely predicts the drift vector given the momentum perturbation.

Finally, it should be noted that the drift phenomena may have some practical implications in the design of devices (such as spacecraft) with interacting multiple rotating parts. Namely, as has already been mentioned, if a stable platform is required, then formal stability is inadequate, and the regime of zero total angular momentum should be avoided. Conversely, if reorientation is required, then a formally stable relative equilibria with zero total angular momentum may be desirable. In that case, a controlled slow reorientation could be accomplished using drift induced by a small perturbation to nonzero total angular momentum.

2 Two Coupled Rigid Bodies

We begin by recalling the Lagrangian description for the free motion of two rigid bodies coupled by an ideal spherical joint. For more details see Grossman, Krishnaprasad and Marsden [1988], Patrick [1989], and Patrick [1991]. To obtain the configuration space of this system, fix an inertial frame, and let ρ_1 and ρ_2 be two distributions on $\mathbb{R}^3$, representing the densities of the two bodies in some reference configuration. Without loss of generality, assume that the spherical joint in this configuration is located at the origin. Then take as configuration space the manifold $SO(3)^2 \times \mathbb{R}^3$: a point $q \in \mathbb{R}^3$ on body $i \in \{1,2\}$ in the reference configuration has position $A_i q + w$ in the configuration (A_1, A_2, w). In a state described by the tangent vector $(A_i, w, \dot{A}_i, \dot{w})$, the kinetic energy is just

$$KE = \frac{1}{2}\int |\dot{A}_1 q + \dot{w}|^2 \rho_1(q)\, dq + \frac{1}{2}\int |\dot{A}_2 q + \dot{w}|^2 \rho_2(q)\, dq,$$

and since there is no potential, this function is also the Lagrangian. A reduction corresponding to the action of the group $\mathbb{R}^3$ by addition in the w variable can then be performed, a result that may also be obtained by moving to the center a mass frame, and yields a Lagrangian system on the configuration space $Q = SO(3)^2$. After this reduction, the remaining symmetry is that of left diagonal multiplication of $SO(3)$ on the configuration space $SO(3)^2$.

We now specialize to the identical body axially symmetric case, where the two bodies are identical and axially symmetric, and the bodies are joined along their axes of symmetry. With these assumptions, the continuous symmetry group rises to $SO(3) \times S^1 \times S^1$, which acts on the configuration space $SO(3)^2$ by

$$(B, \theta_1, \theta_2) \cdot (A_1, A_2) = \left(BA_1 \exp(-\theta_1 \boldsymbol{k}^\wedge), BA_2 \exp(-\theta_2 \boldsymbol{k}^\wedge)\right),$$

where $a \in \mathbb{R}^3 \mapsto a^\wedge \in \mathfrak{so}(3)$ is the usual identification of $\mathbb{R}^3$ and $\mathfrak{so}(3)$ by

$$a^\wedge = \begin{bmatrix} 0 & -a^3 & a^2 \\ a^3 & 0 & -a^1 \\ -a^2 & a^1 & 0 \end{bmatrix}.$$

The number of essential parameters reduces to two: a measure of "thickness", α, and a measure of the degree of coupling between the bodies, β, which is zero if the joint lies at the mutual centers of mass of the bodies. With the identification of the tangent bundle of $SO(3)^2$ with $SO(3)^2 \times (\mathbb{R}^3)^2$ by left translation, the Lagrangian takes the form

$$L = \frac{1}{2} \begin{bmatrix} \Omega_1{}^t & \Omega_2{}^t \end{bmatrix} J(A) \begin{bmatrix} \Omega_1 \\ \Omega_2 \end{bmatrix},$$

where $A = [A^{ij}] = A_1{}^t A_2$ and

$$J(A) = \begin{bmatrix} 1 & 0 & 0 & -\beta A^{22} & \beta A^{21} & 0 \\ 0 & 1 & 0 & \beta A^{12} & -\beta A^{11} & 0 \\ 0 & 0 & \alpha & 0 & 0 & 0 \\ -\beta A^{22} & \beta A^{12} & 0 & 1 & 0 & 0 \\ \beta A^{21} & -\beta A^{11} & 0 & 0 & 1 & 0 \\ 0 & 0 & 0 & 0 & 0 & \alpha \end{bmatrix}.$$

This is the Lagrangian system known as two axially symmetric rigid bodies coupled be a ball and socket joint, or 2ACRB, for short.

All relative equilibria for 2ACRB are explicitly known, as are their formal stability; a complete list can be found in Patrick [1991]. The relative equilibria and their properties that are of interest here are as follows:

Theorem 2 *For typographical convenience, define the coefficient*

$$\kappa_\gamma^{t_1 t_2 t_3} = \frac{(t_1 \cos t_3 - t_2)(\beta t_2 - \gamma t_1)}{\alpha t_1 \sin t_3}.$$

Then, for every triple $t_1, t_2, t_3 \in \mathbb{R}$ such that $t_1 \neq 0, t_2 \neq 0, 0 < t_3 < \pi$, the point $\big(Id, \exp(\theta \boldsymbol{j}^\wedge), \Omega_1, \Omega_2\big)$ calculated by

$$\theta = t_3, \Omega_1 = t_1 \boldsymbol{i} - \kappa_1^{t_1 t_2 t_3} \boldsymbol{k}, \Omega_2 = t_2 \boldsymbol{i} + \kappa_1^{t_2 t_1 t_3} \boldsymbol{k}$$

is a relative equilibrium of 2ACRB with Lie algebra element $(\Omega, \sigma_1, \sigma_2) \in \mathfrak{so}(3) \times \mathbb{R}^2$ given by

$$\sigma_1 = \kappa_{1-\alpha}^{t_1 t_2 t_3}, \sigma_2 = -\kappa_{1-\alpha}^{t_2 t_1 t_3}, \Omega = t_1 \boldsymbol{i} + \frac{t_1 \cos t_3 - t_2}{\sin t_3} \boldsymbol{k}.$$

The magnitude of the total angular momentum of this relative equilibrium is

$$\frac{t_2{}^2(\beta t_2^2 - 2t_1 t_2 + \beta t_1^2)^2 (t_2^2 - 2 t_1 t_2 \cos t_3 + t_1^2)}{t_1^2 t_2^2 \sin^2 t_3}.$$

This relative equilibrium is formally stable if and only if

$$\beta t_1^2 - 2t_1 t_2 + \beta t_2^2 \geq 0.$$

In particular, the nongeneric relative equilibria in this class are just those with zero total angular momentum; that is, those relative equilibria arising from triples (t_1, t_2, t_3) such that

$$\beta t_2^2 - 2t_1 t_2 + \beta t_1^2 = 0.$$

In this state, the system undergoes a nontrivial overall rotation while each body spins in such a way that the joint angle is fixed and the total angular momentum is zero.

3 The Numerical Calculation of Drift

In the general context of a Hamiltonian system with symmetry (P, ω, G, H), the numerical computation of the drift begins with the numerical simulation of the system: One chooses an initial state p_0 to be a small perturbation of one relative equilibrium, say p_e, with momentum μ_e and associated Lie algebra element ξ_e. The output of the numerical simulation is a sequence of points of phase space, say

$$p_0, p_1 \cdots, p_n \in P,$$

where p_i is approximately the state of the system at time $i\Delta t$ when that system is in state p_0 at time $t = 0$, and Δt is small enough the the system moves little over a time interval of length Δt.

By theorem (1), the motion starting at p_0 is subsequently near $G_{\mu_e} \cdot p_e$, so elements $g_i \in G_{\mu_e}$ can be found such that p_i is approximated well by $g_i p_0$, by setting g_i to the $g \in G_{\mu_e}$ that minimizes $d(gp_0, p_i)$, where d is some distance function on TQ. A standard nonlinear multidimensional minimization algorithm may be used in conjunction with any convenient distance function. As this minimization is computationally burdensome, one selects a small integer $N_c \approx 20$, and then another integer k, such that $kN_c\Delta t$ is on the order of the smallest period of the quasiperiodic motion $t \mapsto \exp(\xi_e t)$, and then the sequence p_i is sampled every k points. This procedure ensures that the fastest moving system component will span one of its periods every N_c elements in the sequence g_i. Insofar as the the sequence g_i deviates from the closure of the one parameter subgroup $\{\exp(\xi_e t) \mid t \in \mathbb{R}\}$, drift will be detected.

To quantify that deviation of group elements, note that, over a time interval of length $kN_c\Delta t$, if the perturbation of p_0 from p_e is small, the system will be close to a state of relative equilibrium characterized by a Lie algebra element, say η. Thus, one seeks $\eta \in \mathfrak{g}$ such that the equations

$$\exp(jk\Delta t\eta) g_n p_e = g_{n+j} p_e,$$

or equivalently,

$$\exp(jk\Delta t\eta) = g_{n+j} g_n^{-1}, \tag{1}$$

for $0 \leq j \leq N_c - 1$, are satisfied as closely as possible. Iteratively applied, over successive time intervals of length $kN_c\Delta t$, this will result in a sequence of $L = \lfloor n/kN_c \rfloor \ll n$ Lie algebra elements η_l, $1 \leq l \leq L$, each representing the evolution of the system over a moderate time interval $kN_c\Delta t$ beginning at time $(l-1)kN_c\Delta t$. By G invariance, the evolution at $g_n p_e$ is exactly

$$g_n \exp(\xi_e t) p_e = g_n \exp(\xi_e t) g_n^{-1} g p_e = \exp(g_n \xi_e t) g_n p_e,$$

so that the η_j all are near the adjoint orbit of ξ_e.

The finding of the Lie algebra elements η_j by the fitting of the η_j to the data $g_{n+j}g_n^{-1}$ through equation (1) depends on the group under consideration. Returning to the system 2ACRB, and redefining Δt to be $k\Delta t$, the problem is reduced to the following: given

$$(A_j, \theta_j^1, \theta_j^2) \in SO(3) \times (S^1)^2, \quad 0 \le j \le N_c - 1$$

find $\Omega \in \mathbb{R}^3$ and $\sigma^1, \sigma^2 \in \mathbb{R}$ such that the equations

$$A_j = \exp(j\Delta t \Omega^\wedge), \quad \theta_j^1 = j\Delta t\sigma^1, \quad \theta_j^2 = j\Delta t\sigma^2, \tag{2}$$

are satisfied as closely as possible. In fact, this is quite easy for the second and third of equations (2): the second equation simply requires the fitting of the points $(j\Delta t, \theta_j^1)$ to the line $\theta = \sigma^1 t$, and similarly with the third, and this is a standard least squares problem:

$$\sigma^1 = \frac{\sum_{j=0}^{N_c-1} j\Delta t\theta_i^1}{\sum_{j=0}^{N_c-1} (j\Delta t)^2}, \quad \sigma^2 = \frac{\sum_{j=0}^{N_c-1} j\Delta t\theta_i^2}{\sum_{j=0}^{N_c-1} (j\Delta t)^2}.$$

The finding of an Ω so that the first of equations (2) is satisfied in some least squares sense can be accomplished as follows: note that the elements $\exp(j\Delta t\eta)$ are all right handed rotations about the unit vector in direction Ω. Thus, setting

$$A = \frac{1}{N_c - 1} \sum_{j=0}^{j=N_c-1} (A_j + {A_j}^t),$$

one finds $\pm\Omega/|\Omega|$ by minimizing over unit vectors $x \in \mathbb{R}^3$ the expression

$$\begin{aligned} \sum_{j=0}^{j=N_c-1} |A_j x - x|^2 &= \sum_{j=0}^{j=N_c-1} 2|x|^2 - x^T(A_j + {A_j}^t)x \\ &= (N_c - 1)(2 - x^t A x). \end{aligned}$$

The minimum occurs when x is one of the two unit eigenvectors of A with largest eigenvalue, say λ, and then the root mean square error is $\sqrt{2-\lambda}$. Fixing one of these unit eigenvectors, say x_λ, one completes it to a right handed orthonormal basis of $\mathbb{R}^3$, say x_1, x_2, x_λ, and then the two real numbers $(A_j x_1 \cdot x_1, A_j x_1 \cdot x_2)$ will closely move in a circle with angular velocity $\sigma\Delta t$, where $\Omega = \sigma x_\lambda$. Then σ is found by lifting this path in the circle to the universal covering $\mathbb{R}$, and then fitting the resulting angles θ_j to a straight line, as in the computation of σ_1 and σ_2 above.

4 Observations

In this section we discuss a perturbation of the particular relative equilibrium obtained by setting

$$\alpha = \frac{1}{2}, \quad t_1 = 1, \quad t_2 = 4, \quad t_3 = \frac{\pi}{3} \tag{3}$$

in theorem (1). This relative equilibrium will have total angular momentum zero if and only if

$$\beta t_2^2 - 2t_1t_2 + \beta t_1^2 = 0,$$

so we take

$$\beta = \frac{2t_1 t_2}{t_1^2 + t_2^2} = \frac{8}{17},$$

and then $\xi_e = (\tilde{\xi}_e, \sigma_e^1, \sigma_e^2)$, where

$$\tilde{\xi}_e = \left[\begin{array}{ccc} 1 & 0 & -\frac{7}{3}\sqrt{3} \end{array}\right] \tag{4}$$

and

$$\sigma_e^1 = -\frac{329}{51}\sqrt{3}, \quad \sigma_e^2 = \frac{26}{51}\sqrt{3}. \tag{5}$$

Specified momentum perturbations of a relative equilibrium with zero total angular momentum can be achieved as follows. Suppose one is given a momentum perturbation

$$\Delta\mu = (\Delta\tilde{\mu}, \Delta\mu_1, \Delta\mu_2) \in \mathfrak{so}(3) \times \mathbb{R}^2,$$

a perturbation of the joint angle $\Delta\theta$, and another small value $\Delta\pi_2^2$. For an element of phase space $(A_1, A_2, \Omega_1, \Omega_2)$, set

$$\left[\begin{array}{c} \pi_1 \\ \pi_2 \end{array}\right] = J(A) \left[\begin{array}{c} \Omega_1 \\ \Omega_2 \end{array}\right],$$

which is exactly the Legendre transformation with identifications by left translation. The momentum is then

$$(\tilde{\mu}, \mu_1, \mu_2) = (\pi_1 + A\pi_2, -\pi_1 \cdot \boldsymbol{k}, -\pi_2 \cdot \boldsymbol{k}).$$

If $A_1 = \mathrm{Id}$ and $A_2 = \exp(\theta \boldsymbol{j}^\wedge)$, one seeks $\Delta\pi_1$ and $\Delta\pi_2$ such that the equations

$$\pi_1 + \Delta\pi_1 + \exp(\theta' \boldsymbol{j}^\wedge)(\pi_2 + \Delta\pi_2) = \Delta\tilde{\mu}, \tag{6}$$
$$-(\pi_1 + \Delta\pi_1) \cdot \boldsymbol{k} = \mu_1 + \Delta\mu_1, \quad -(\pi_2 + \Delta\pi_2) \cdot \boldsymbol{k} = \mu_1 + \Delta\mu_2, \tag{7}$$

are satisfied, where $\theta' = \theta + \Delta\theta$. Since $-\boldsymbol{k} \cdot \pi_1 = \mu_1$ and $-\boldsymbol{k} \cdot \pi_2 = \mu_2$, equations (7) are equivalent to the equations

$$\Delta\pi_1 \cdot \boldsymbol{k} = -\Delta\mu_1, \quad \Delta\pi_2 \cdot \boldsymbol{k} = -\Delta\mu_2,$$

which simply determine the third components of π_1 and π_2. For the other components, one solves (6) to obtain

$$\Delta\pi_2^1 = \frac{\Delta\pi_1^3 + \Delta\pi_2^3 \cos\theta' - a^3}{\sin\theta'},$$

$$\Delta\pi_1^1 = a^1 - \Delta\pi_2^3 \sin\theta' - \Delta\pi_2^1 \cos\theta',$$

$$\Delta\pi_1^2 = a^2 - \Delta\pi_2^2,$$

where $a \in \mathbb{R}^3$ is

$$a = \Delta\tilde{\mu} - \pi_1 - \exp(\theta' \boldsymbol{j}^\wedge)\pi_2.$$

Finally, the initial data in tangent phase space that has the specified momentum perturbation is $\left(\mathrm{Id}, \exp(\theta' \boldsymbol{j}^\wedge), \Omega_1', \Omega_2'\right)$, where

$$\begin{bmatrix} \Omega_1' \\ \Omega_2' \end{bmatrix} = J\left(\exp(\theta' \boldsymbol{j}^\wedge)\right)^{-1} \begin{bmatrix} \pi_1 + \Delta\pi_1 \\ \pi_2 + \Delta\pi_2 \end{bmatrix}.$$

Of course, there are many ways to perturb any relative equilibrium, and in fact, 6 more values are required to uniquely determine a perturbation, since the dimension of the phase space is 12.

As discussed above, it is the slow evolution of the Lie algebra elements η_l that is of interest. For the system 2ACRB, this simplifies somewhat, because that evolution occurs near the adjoint orbit of ξ_e, and the adjoint action fixes the two S^1 parts of any element in the Lie algebra of $SO(3) \times (S^1)^2$. Thus, if we set

$$\eta_l = (\tilde{\eta}_l, \sigma_l^1, \sigma_l^2) \in \mathfrak{so}(3) \times \mathbb{R}^2,$$

then σ_l^1 and σ_l^2 will not vary over large times and $\tilde{\eta}_l$ will evolve near a sphere of radius $|\tilde{\xi}_e|$. When observing this perturbation over a long time, the system evolves thusly: the the bodies spin, there is some overall rotation, the joint angle moves only slightly, and the overall rotation slowly changes direction.

As an example of how the overall rotation evolves, figure (1) displays the vectors η_l when the perturbation is

$$\Delta\mu = \begin{bmatrix} .05 & .05 & .1 \end{bmatrix}, \quad \Delta\theta = .025, \quad \Delta\pi_1^2 = 0. \tag{8}$$

Observe that the vectors η_l lie nearly in a plane, which is shown as the shaded region. As well, the η_l lie near a circle with center on that plane's normal line through the origin, so the η_l lie near to the circle in the sphere $|\eta| = |\tilde{\xi}_e|$ that contains $\tilde{\xi}_e$. In addition, figure (2) shows the time dependence of the angle ζ that the vector η_l makes with the vector ξ_e, measured counter clockwise in the figure, so that it is apparent that the motion of the η_l is uniformly circular. As well, the angular velocity of the circular motion is .00878, so that the system rotates approximately

$$\frac{|\tilde{\xi}_e|}{.00878} = 474,$$

times due to the overall rotation vector $\tilde{\xi}_e$ for every single rotation due to drift, while the bodies rotate approximately

$$\frac{|\sigma_e^1|}{.00878} = 1298, \quad \frac{|\sigma_e^2|}{.00878} = 101$$

times due to the spin rates σ_e^1 and σ_e^2 respectively for every single rotation due to drift. We summarize these data in the following observation:

Observation 1 *Perturbations of the relative equilibrium (3) cause precession: the vectors $\tilde{\eta}_l$ move slowly and uniformly near a circle that contains $\tilde{\xi}_e$ and lies in the sphere $|\tilde{\eta}| = |\tilde{\xi}_e|$.*

When examining figure (1), it should be remembered that the motion depicted there is caused by a perturbation of a relative equilibrium. The finest motion in the figure is associated to a sampling of an apparently quasiperiodic motion that is only near to a relative equilibrium, and that includes, for example, near linear modes at the corresponding equilibrium of the reduced phase space. The less fine humped structure in the figure is caused by a slow rotation of approximate planes defined by sets of vectors η_l spanning times that are only moderate compared to Δt. At the time of writing, this humped structure has not been investigated, but the deviations in η_l that produce those humps could be artifacts of the choice of the distance function used to compute the group elements g_i. However, the main feature of the figure, and the feature that is associated to the long term dynamics near the relative equilibrium, is that the average motion of the vectors $\tilde{\eta}_l$ is that of uniform rotation is the way described in observation (1).

In view of observation (1), the evolution of the η_l may characterized by a single rotation vector $\eta_{\mathrm{drift}} \in \mathbb{R}^3 = \mathfrak{so}(3)$. One calculates η_{drift} so that the equation

$$\exp\left(lN_c\Delta t\eta_{\mathrm{drift}}^{\wedge}\right)\eta_0 = \eta_l$$

is satisfied in some least squares sense as follows: first compute the average

$$\bar{\eta} = \frac{1}{L}\sum_{l=1}^{L}\eta_l,$$

and then set

$$B = \frac{1}{L}\sum_{l=0}^{L}(\eta_l - \bar{\eta})(\eta_l - \bar{\eta})^t.$$

Then one finds a unit vector orthogonal to the plane of rotation by minimizing over unit vectors $y \in \mathbb{R}^3$ the expression

$$\begin{aligned}\sum_{l=1}^{L}|y\cdot(\eta_l - \bar{\eta})|^2 &= \sum_{l=1}^{L}y^t(\eta_l - \bar{\eta})(\eta_l - \bar{\eta})^t y\\ &= Ly^tBy.\end{aligned}$$

The minimum occurs when y is one of the two unit eigenvectors of B, say y_ρ, with smallest eigenvalue ρ, and then the root mean square error is $\sqrt{\rho}$. By completing y_ρ to a right handed orthonormal basis of $\mathbb{R}^3$, one can determine the rotation rate in exactly the way that the elements η_l were themselves calculated.

Now that the drift has been quantified, we can attend to a natural worry: since the simulator time step must be small compared to the fastest motion of the system, and since the drift is slow in that time scale, the simulator must be run over long times compared to the time step, so the possibility exists that the observed drift is an artifact arising from numerical computation. In fact, a standard test can be performed: the drift rotation vector is calculated for some initial data and then compared with the same vector calculated using the same initial data but at half the time step and the same total time. For example, with the initial data implied by the perturbation defined by (8), and with time step .005 and 30,000 iterations, the drift rotation vector was calculated to be

$$\begin{bmatrix} -0.00353278 & -0.0036587 & -0.00720471 \end{bmatrix}, \tag{9}$$

while at time step .0025 and 60,000 iterations the rotation vector was calculated to be

$$\begin{bmatrix} -0.00353278 & -0.00365869 & -0.00720471. \end{bmatrix}, \tag{10}$$

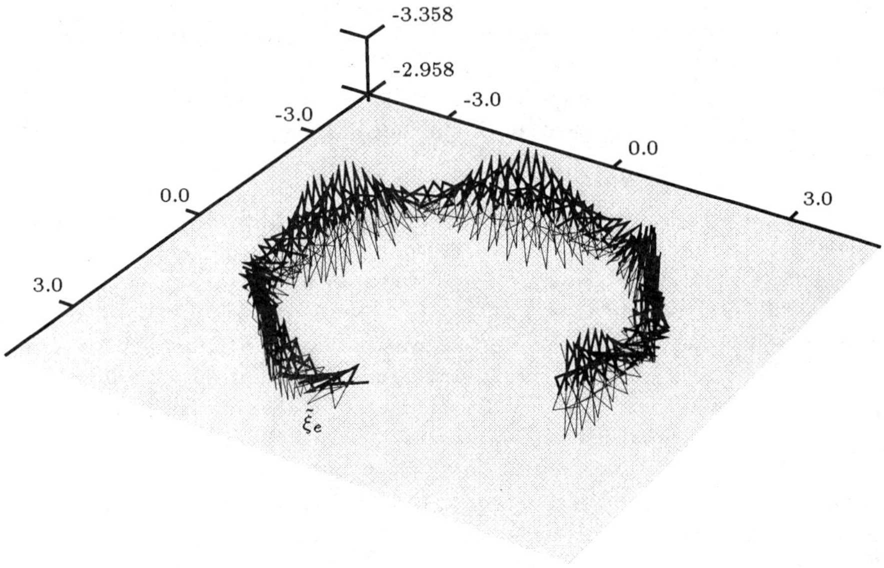

Figure 1 The rotation vectors η_l.

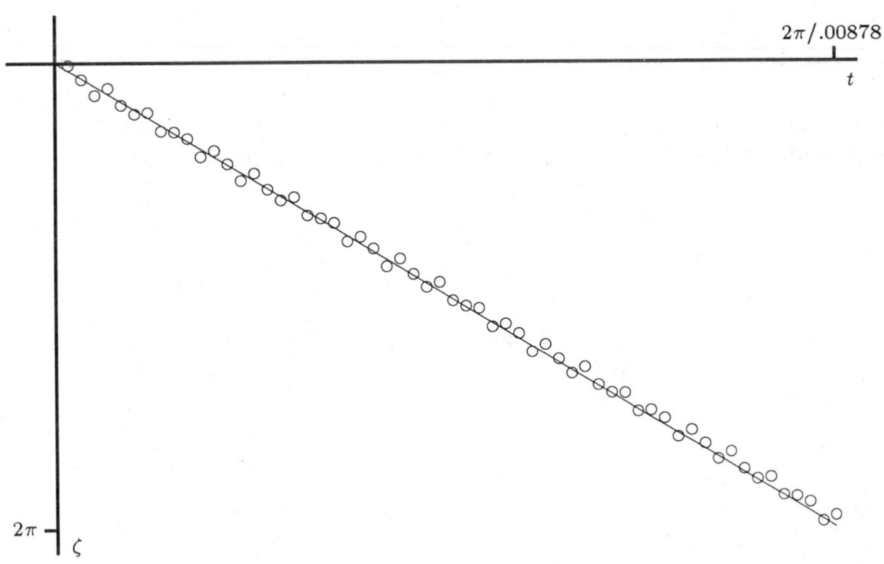

Figure 2 The time evolution of the angle ζ between η_l and $\tilde{\xi}_e$.

which agrees well with (9). Although any numerical scheme would suffice, the numerical scheme used to generate the 2ACRB simulations for this article was the third order momentum preserving symplectic integration algorithm found in Patrick [1991] and Patrick [1993].

We now turn to the dependence of the drift vector on the perturbation. As an example of this, figure (3) displays the magnitude of the drift vector at perturbations with

$$\Delta\tilde{\mu} = \nu \begin{bmatrix} .1 & .1 & .2 \end{bmatrix}, \quad \Delta\theta = 0, \quad \Delta\pi_1^2 = 0,$$

as ν varies from 0 to 1. Also shown is the linear approximation at the origin to that graph. As well, the root mean square measure of the deviation from the mean of the corresponding normalized drift vectors was only .037, so the drift vectors have nearly constant direction, and as the data in figure (3) deviates little from the linear approximation shown there, a linear dependence of η_{drift} on ν is clear, at least for small ν. Figure (4) displays the normalized drift vectors for perturbations with ν as above but for varying $\Delta\theta$. As the lines displayed there are nearly flat (and in become more flat as ν vanishes), it appears that the drift vector is independent of $\Delta\theta$. Similar results could be obtained for other types of constant $\Delta\tilde{\mu}$ perturbations. Thus, given the 12 dimensional continuum of possible perturbations, the following observation is remarkable:

Observation 2 *The drift vector η_{drift} nearly depends linearly on the momentum perturbation $\Delta\tilde{\mu}$, and depends only on $\Delta\tilde{\mu}$.*

Observation (2) can be formalized as follows: there is a linear *drift map* $\tilde{\Xi}$: $\mathfrak{so}(3)^\star \to \mathfrak{so}(3)$ such that

$$\eta_{\text{drift}} = \tilde{\Xi}_{p_e} \cdot dJ(p_e)\delta v$$

Here the perturbation implied by $\delta v \in T_{p_e}P$ is $p_e + \delta v$ in any coordinate chart at p_e. This drift map can calculated numerically simply by calculating the drift vectors corresponding to perturbations parallel to the standard basis vectors $\boldsymbol{i}, \boldsymbol{j}, \boldsymbol{k}$ in $\mathfrak{so}(3)$ and $\mathfrak{so}(3)^\star$. Doing this for the relative equilibria (3) yields

$$\begin{aligned} \tilde{\Xi}_{p_e} &= \begin{bmatrix} -0.00757 & 7.38 \times 10^{-6} & 2.27 \times 10^{-5} \\ -6.96 \times 10^{-7} & -0.00759 & 1.42 \times 10^{-8} \\ 4.78 \times 10^{-5} & -2.14 \times 10^{-5} & -0.0078 \end{bmatrix} \\ &= -.07650 \begin{bmatrix} .989 & 0 & 0 \\ 0 & .991 & 0 \\ 0 & 0 & 1.019 \end{bmatrix} \\ &\quad + 10^{-5} \begin{bmatrix} 0 & 0.738 & 2.27 \\ -0.0696 & 0 & .00142 \\ 4.78 & -2.14 & 0 \end{bmatrix} \end{aligned} \tag{11}$$

In particular, it is apparent that the drift map is nearly a constant multiple of the identity matrix, so that the direction of the drift vector is parallel or opposite to the momentum perturbation $\Delta\tilde{\mu}$. In this case, it is opposite, and thus we have the following observation:

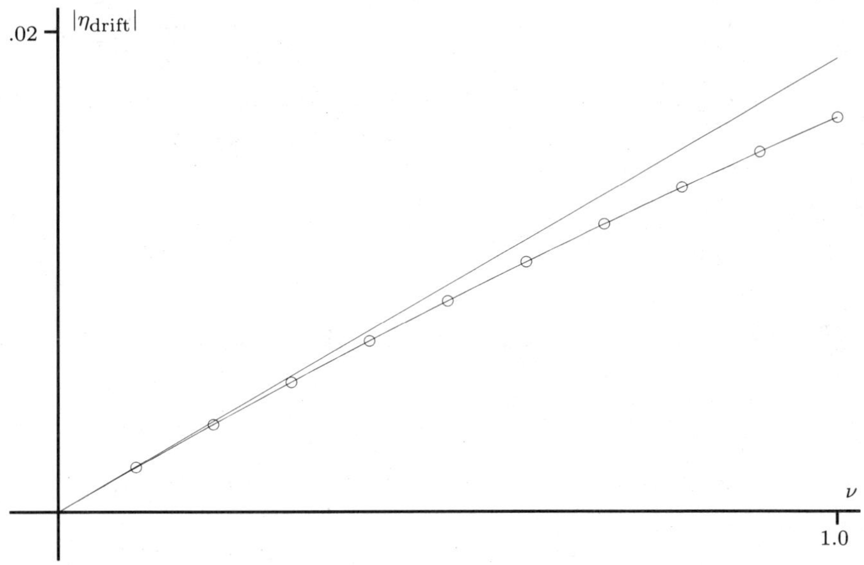

Figure 3 The magnitude of the drift vector is nearly linear in $\Delta\tilde{\mu}$.

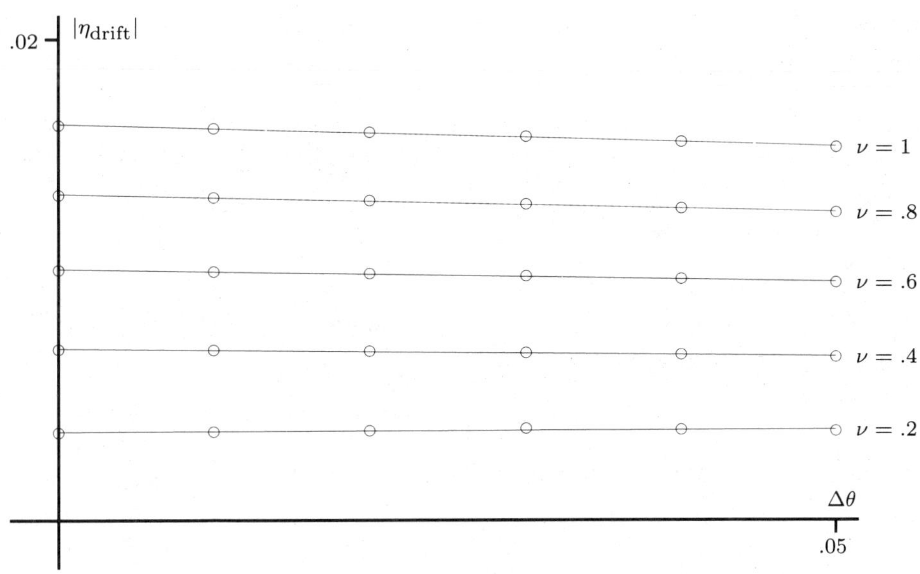

Figure 4 The magnitude of the drift vector does not depend on perturbations $\Delta\theta$ of the joint angle.

Observation 3 *The rotation vector η_{drift} of the drift and the momentum perturbation $\Delta\tilde{\mu}$ are parallel or opposite. The drift* can *be retrograde; that is, a perturbation with momentum $\Delta\tilde{\mu}$ can drift with rotation vector directed oppositely to $\Delta\tilde{\mu}$.*

As can be seen, then, the main theoretical problem posed by the drift phenomena is the calculation of the drift map $\tilde{\Xi}_{p_e}$, and the explication of observations (1) through (3). The following discussion is indicative of what may be involved: The Hamiltonian vector field at p_e admits a linearization, of a sort, by noting that the vector field

$$X_{H_{\xi_e}} = X_H - \xi_e$$

vanishes exactly at p_e, by definition. The linearization of this vector field is a linear operator on $T_{p_e}P$, and can be defined by

$$dX_{H_{\xi_e}} \cdot V(p_e) = [V, X_{H_{\xi_e}}](p_e),$$

where V is a vector field on TP and $[\ ,\ ]$ is the Lie bracket of vector fields. Now, the characteristic polynomial of that operator is exactly

$$\frac{1}{63}x^6\left(3x^2+52\right)^2\left(7x^2+400\right),$$

the roots of which are purely complex or zero, so no drift is apparent from the eigenvalues alone. However, the minimum polynomial of the linearization is

$$\frac{1}{63}x^2\left(3x^2+52\right)^2\left(7x^2+400\right)$$

which is reducible, so the Jordan decomposition of this operator has a nontrivial nilpotent part (see Hoffman and Kunze [1971]), which implies instability in the linear evolution.

Guessing that the nilpotent part $\bar{N}$ of the linearization is responsible for the drift phenomena, one can compute that operator. In addition to the fact that $\bar{N}$ is nilpotent of order 2, one may then verify the following two facts:

1. $\bar{N}$ has image in $T_{p_e}(G \cdot p_e)$.
2. $\bar{N}$ is zero on $\ker dJ(p_e)$.

As a consequence of these two facts, $\bar{N}$ determines a linear map N from $\mathfrak{so}(3)^\star \times \mathbb{R}^2$ to $\mathfrak{so}(3) \times \mathbb{R}^2$ which satisfies

$$(N \cdot dJ(p_e)v) \cdot p_e = \bar{N} \cdot v,$$

for all $v \in T_{p_e}P$. With respect to standard bases, the matrix of N is

$$N = \begin{bmatrix} \frac{2839}{2119260} & 0 & \frac{-1133713\sqrt{3}}{6357780} & \frac{-476\sqrt{3}}{2925} & \frac{-17\sqrt{3}}{5850} \\ 0 & \frac{-204}{2717} & 0 & 0 & 0 \\ \frac{-1133713\sqrt{3}}{6357780} & 0 & \frac{7458631}{6357780} & \frac{3332}{2925} & \frac{119}{5850} \\ \frac{-476\sqrt{3}}{2925} & 0 & \frac{3332}{2925} & \frac{47}{15} & 0 \\ \frac{-17\sqrt{3}}{5850} & 0 & \frac{119}{5850} & 0 & \frac{13}{15} \end{bmatrix},$$

which has characteristic polynomial proportional to p_N, where

$$p_N = \left(658125\,x^3 - 3454875\,x^2 + 4170082\,x - 1446836\right)\left(2717\,x + 204\right)^2.$$

Interestingly,

$$-\frac{204}{2717} = -.07508,$$

a value that predicts the drift matrix, since it deviates from the multiplier of the identity in (11) by only 1.9%.

Some other entries of N correspond to drift that is *time-like*. For example, by multiplying N into the momentum perturbation, the last two rows of N predict the constant values of the last two components (that is, the components corresponding to $(S^1)^2$) of the Lie algebra elements η_l. Noteworthy as well is the fact that the last two columns of N are vectors antiparallel to $\tilde{\xi}_e$, so that N predicts that momentum perturbations with $\Delta\tilde{\mu} = 0$ but $\Delta\mu_1 \neq 0$ or $\Delta\mu_2 \neq 0$ yield zero nontrivial drift. Unfortunately, at the time of writing, the expected time-like drift along ξ_e induced, for example, by perturbations with only nonzero $\Delta\mu_1$, has not been related to the matrix N. In any case, ample justification has been provided for the following observation:

Observation 4 *The nilpotent part of the Jordan decomposition of linearization of the evolution vector field at a relative equilibrium induces a symmetric operator from the dual of the Lie algebra of the symmetry group to that Lie algebra. Drift, both time-like and nontrivial, is related to that operator.*

Presumably, the matrix N corresponds to or constructs a vector field on the group orbit of p_e that approximates the exact evolution vector field X_H, and that approximation can be computed by linearization and averaging of X_H along the *whole group orbit* of p_e. This matter, and other related theoretical investigations, will be subjects of the forthcoming article Patrick [1995].

References

Abraham, R. and Marsden, J. E. [1978], *Foundations of Mechanics,* Second Edition. Addison-Wesley, Reading, Mass.

Channell, P. J. and Scovel, C. [1990], Symplectic integration of Hamiltonian systems. *Nonlinearity,* **3**, 231–259.

Grossman, R., Krishnaprasad, K. S., and Marsden, J. E. [1988], *The dynamics of two coupled rigid bodies,* In: Dynamical Systems Approaches to Nonlinear Problems in Systems and Circuits (Levi, M. and Salam, F., ed.), 373–378.

Hoffman, K. and Kunze, R. [1971], *Linear Algebra,* Second Edition, Prentice-Hall, Englewood Cliffs, New Jersey.

Holm, D., Marsden, J. E., Ratiu, T. and Weinstein, A. [1984], *Stability of rigid body motion using the energy-Casimir method,* In: Fluids and Plasmas: Geometry and Dynamics (Marsden, J.E., ed.), American Mathematical Society, Providence, R.I.

Marsden, J. E., Simo, J. C., Lewis, D., and Posbergh, T. A. [1989], *Block diagonalization and the energy-momentum method,* In: Dynamics and Control of

Multibody Systems (Marsden, J.E., Krishnaprasad, P.S., and Simo, J.C., ed.), American Mathematical Society, Providence, R.I., 315–335.

Patrick, G. W. [1989], *The dynamics of two coupled rigid bodies in three space,* In: Dynamics and Control of Multibody Systems (Marsden, J.E., Krishnaprasad, P.S., and Simo, J.C., ed.), American Mathematical Society, Providence, R.I., 297–313.

Patrick, G. W. [1991], *Two axially symmetric coupled rigid bodies: relative equilibria, stability, bifurcations, and a momentum preserving symplectic integrator,* PhD thesis, University of California at Berkeley.

Patrick, G. W. [1992], *Relative equilibria in Hamiltonian systems: The dynamic interpretation of nonlinear stability on the reduced phase space,* Journal Geometry Physics, **9**, 111–119.

Patrick, G. W. [1993], *A vectorizing momentum preserving symplectic integrator,* In: Fields Institute Communications (Enos, M.J., ed.), **1**, 247–261.

Patrick, G. W. [1995], *Relative equilibria of Hamiltonian systems with symmetry: linearizations, smoothness and drift,* Journal Nonlinear Science, to appear.

Simo, J. C., Lewis, D., and Marsden, J. E. [1991], *Stability of relative equilibria. Part I: The reduced energy-momentum method,* Archive Rational Mechanics Analysis, **115**, 15–59.

Fields Institute Communications
Volume 8, 1996

The Dynamics of Reversible Systems and Spatio-temporal Complexity in Biology

B. D. Sleeman
Department of Mathematics and Computer Science
University of Dundee
Dundee DD1 4HN, Scotland UK
Current Address: School of Mathematics
University of Leeds
Leeds LS2 9JT, UK

1 Introduction

The purpose of this paper is to demonstrate that complex spatial and temporal structural behaviour observed in a wide variety of biological phenomena can be understood through the study of Hamiltonian systems and dynamical systems theory in general. It is a primary goal to delineate those characteristics of biological behaviour which result from purely evolutionary processes and subsequent dispersal. In so doing we draw attention to a number of open problems associated with dynamical systems and simultaneously indicate a new area of application of Hamiltonian systems.

The basic, or generic model to be discussed is of the form

$$u_{n+1} = Q[u_n](x), \tag{1.1}$$

where $u_n(x)$ may be thought of as a population density at time $n\tau$ (τ-life time) at the position $x \in \mathbb{R}$ of a one-dimensional habitat. The operator Q, defined on the habitat, is typically one of the forms

(i) Reaction-diffusion operator,
(ii) Integral operator,
(iii) Partial difference operator.

Rather than describe the wide variety of biological and physiological phenomena which can be modeled by (1.1) (c.f. Sleeman [1989a], [in preparation]) we shall concentrate on two subject areas, namely

1. Population genetics.

2. Ecological.

1991 *Mathematics Subject Classification.* Primary: 58F05, 58F13; Secondary: 92D10, 92D40.

Where appropriate we shall mention other models wherein new and exciting developments are taking place and deserve greater mathematical attention. These include certain classes of automata models which are particularly useful in attempting to model strategic rather than random dispersal of species.

2 Population Genetics

A gene of a certain species occurs in two forms called alleles denoted by A and a. Each individual of the species has two alleles at the chromosone locus corresponding to the gene in question, each of which may be of type A or type a. Thus there are three genotypes denoted by AA, Aa and aa each distributed with population densities $\rho_{AA}(x)$, $\rho_{Aa}(x)$ and $\rho_{aa}(x)$ respectively.

If $u(x)$ represents the fraction of allele A in the parent population at x then

$$u(x) = \frac{2\rho_{AA} + \rho_{Aa}}{2(\rho_{AA} + \rho_{Aa} + \rho_{aa})}. \tag{2.1}$$

Under the assumption of the Hardy-Weinberg law, with $\rho_{AA}+\rho_{Aa}+\rho_{aa}$ independent of x and non-overlapping generations, the gene fraction $u_n(x)$ at time $n\tau$ is related to $u_{n+1}(x)$ by

$$u_{n+1} = \int_{\mathbb{R}} K(x-y)g(u_n(y))dy \tag{2.2}$$

where K is a probability kernel and g is a growth law. For example we may specify logistic growth to obtain the model

$$u_{n+1} = \int_{\mathbb{R}} e^{-\alpha|x-y|}\beta u_n(y)(1-u_n(y))dy. \tag{2.3}$$

This model will be explored further in the ecological models considered below. We may also develop the precursor of Fisher's classic equation

$$\frac{u_{n+1}-u_n}{\tau} = D\frac{\partial^2 u_n}{\partial x^2} + f(u_n) \tag{2.4}$$

in the case where each genotype has a survival rate close to 1.

Having made the basic assumption that the entire population reproduces and dies out at the end of each time interval τ we cannot let $\tau \to 0$ to derive the classic reaction-diffusion model

$$u_t = D\frac{\partial^2 u}{\partial x^2} + f(u), \tag{2.5}$$

in this situation.

Suppose that the habitat is divided into discrete regions or niches. The population which lives in any one niche during the non-migratory part of the life cycle is called a deme. If, as before, we assume non-overlapping generations and that members of different demes do not mix or interact during a brief period of migration at the end of the life cycle and that mating is random then we derive a model of the form

$$u_{n+1}(j) = \sum_i m_{ji} g_i(u_n(i)), \tag{2.6}$$

where $u_n(i)$ is the gene fraction at generation n located in the ith niche. Again g_i is an assumed growth function.

The model (2.6) has important implications for numerical schemes, Yee, Sweby and Griffiths [1991], used to solve partial differential equations. Thus for example if in (2.4) we select

$$f(u_n) = \beta u_n(1 - u_n) \tag{2.7}$$

and apply the usual second order difference replacement for $\partial^2/\partial x^2$ we have

$$\begin{aligned} u_{n+1}(i) = u_n(i) \quad &+ \quad \frac{\tau D}{h^2}[u_n(i+1) - 2u_n(i) + u_n(i-1)] \\ &+ \quad \tau\beta u_n(i)(1 - u_n(i)), \end{aligned} \tag{2.8}$$

where h is the mesh spacing.

A further observation is that if in (2.4) or (2.8) $D = 0$ then we have purely temporal behaviour and

$$u_{n+1} = u_n + \tau\beta u_n(1 - u_n), \tag{2.9}$$

which is the well known "logistic" model highlighted by May [1976] and demonstrates the familiar "period doubling" route to chaos through a bifurcating sequence of parameters $\tau\beta$.

A basic question to be addressed in this paper is

How do these period-doubling solutions evolve when $D \neq 0$?

In other words we are asked to analyze the interplay between "temporal chaos" and spatial dynamics.

Before leaving the discussion of the model (2.6) which formally has the appearance of a cellular automata model it is appropriate to mention a recently explored model concerning host parasite interactions due to Hassell, Commins and May [1991] which brings out this analogy more closely and in addition offers a further class of biological models which deserves attention from the point of view of dynamical systems theory.

In a homogeneous environment, host-parasite dynamics may be described by

$$\begin{aligned} H_{n+1} &= \lambda H_n f(H_n, P_n), \\ P_{n+1} &= qH_n(1 - f(H_n, P_n)), \end{aligned} \tag{2.10}$$

where H_n represents the host density at generation n and P_n represents the corresponding parasite density. Insect parasitoids lay their eggs on, in or near the bodies of other arthropods and the parasitoid larvae kill the host as they feed on it.

In (2.10), λ is the average number of offspring produced by an unparasitized host; q is the average number of female parasitoids emerging from each host parasitized and $f(H, P)$ is the fraction of hosts escaping parasitism. A common example of the function f is the Bailey-Nicholson form

$$f(H, P) = exp(-aP). \tag{2.11}$$

If parasitoids search randomly and independently then we get the classic Bailey-Nicholson model which is unstable with diverging oscillations. However if we assume that the environment consists of a square grid containing many niches then the dynamics is much more interesting and complex. It evolves in two steps.

I) Apply (2.10) to sub populations to find how many hosts and parasitoids emerge in that niche in the next generation.
II) Allow dispersal by distributing some fixed fraction (μ_H) of hosts and (μ_p) of parasitoids in each niche among other niches according to some specific automata rules. For example we may choose the simplest option and specify that there is equal distribution to the eight nearest neighbours.

Hassel et al. [1991] conducted a number of experiments in which the grid number, as well as the fractions μ_H and μ_p are varied. Depending on the values of these quantities it is found that both the hosts and the parasites distribute themselves over the habitat in a variety of patterns ranging from crystal lattice-like structures, spirals, to chaotic distributions.

Mathematically this system remains to be fully investigated and indeed there are a number of problems to be addressed such as

> What happens if migration strategies are changed ? That is, what are the consequences of altering the cellular automata rules ?

3 Ecology

Consider an organism with synchronous non-overlapping generations. Furthermore suppose there are two distinct stages that define the life cycle of this organism, i.e.

(I) a sedentary stage,
(II) a dispersal stage.

It is assumed that all growth occurs during the sedentary stage (I), and that all migration occurs during the dispersal stage (II) (c.f. Kot [1992]).

In stage (I) the population density can be modeled as

$$N_{t+1} = f(N_t), \tag{3.1}$$

where N_t is the population density at time t and f may take a variety of forms of which the logistic form

$$f(N) = (1+r)N - \frac{r}{K}N^2 \tag{3.2}$$

is typical. Here K is the carrying capacity and r is a growth rate.

To allow for dispersal we introduce a probability kernel $k(x,y)$ which describes the dispersal of N from y to x in the habitat Ω. Stages (I) and (II) can then be combined to give the model

$$N_{t+1}(x) = \int_\Omega k(x,y) f(N_t(y))dy. \tag{3.3}$$

Thus in one spatial dimension we may have

$$N_{t+1}(x) = \int_{-\infty}^{\infty} k(x,y) f(N_t(y))dy, \tag{3.4}$$

where typically k is of convolution type and is given by

$$k(x,y) = \frac{1}{2}\alpha exp(-\alpha|x-y|). \tag{3.5}$$

This particular model should be compared with the model (2.3) arising in population genetics.

In the remainder of this paper we shall provide some insight into the properties of the models (2.4) and (3.4) using ideas drawn from non linear systems theory. In so doing we hope to stimulate interest in a new direction for the application of Hamiltonian formalisms and reversible systems in general. Section 4 will be devoted to partial difference equations motivated by population genetics, leading to a general theory outlined in Section 5, while Section 6 will deal with some questions motivated from ecology.

4 Partial Differential Difference Equations

Consider equations of the form

$$\frac{u_{n+1}-u_n}{\tau} = D\frac{\partial^2 u_n}{\partial x^2} + f(u_n), \tag{4.1}$$

where for the purpose of illustration, $f(u_n)$ has the form

$$f(u) = \beta u(1-u) \tag{4.2}$$

with β a positive number.

In analogy with the well known situation in which $D = 0$, suppose we look for solutions of (4.1) that are period 1 in 'n'. Then we study solutions of

$$D\frac{\partial^2 u_n}{\partial x^2} + f(u_n) = 0, \tag{4.3}$$

i.e.,

$$\begin{aligned} v_{n,x} &= -\frac{1}{D}f(u_n), \\ u_{n,x} &= v_n, \end{aligned} \tag{4.4}$$

which is a Hamiltonian system with Hamiltonian given by

$$H = \frac{1}{2}v_n^2 + \frac{\beta}{D}(\frac{u_n^2}{2} - \frac{u_n^3}{3}). \tag{4.5}$$

In this rather easy case we have a completely integrable system whose dynamics can be fully described by standard phase plane analysis. Moving on to the next step, suppose (4.1) has solutions that are period 2 in 'n'. In this case we look for solutions of the form

$$u_n(x) = q_1(x) + (-1)^n q_2(x). \tag{4.6}$$

As a consequence $q_1(x)$ and $q_2(x)$ satisfy

$$\begin{aligned} Dq_{1,xx} + \beta q_1 - \beta(q_1^2 + q_2^2) &= 0, \\ D\tau q_{2,xx} + (\beta\tau + 2)q_2 - 2\beta\tau q_1 q_2 &= 0. \end{aligned} \tag{4.7}$$

This system is Hamiltonian with two degrees of freedom with H given by

$$H = \frac{1}{2D}(p_1^2 + p_2^2) + \beta\frac{q_1^2}{2} + (\beta + \frac{2}{\tau})\frac{q_2^2}{2} - \beta(\frac{q_1^3}{3} + q_1 q_2^2). \tag{4.8}$$

This Hamiltonian is reminiscent of the classic Henon Heiles system Holmes [1982] and is not known to be completely integrable via the analysis of Ziglin [1982] [1983], except in the limiting case when $\tau \to \infty$. That is it is not known in general whether H has an analytic second integral or not. However despite this seemingly negative result a full analysis of the phase space structure suggests some rather interesting consequences for biology, Mitchell, Murray and Sleeman [1991]. For example if $\epsilon \equiv 1/\tau << 1$ then (4.8) is a perturbation of a completely integrable anti Henon-Heiles type system,

$$H_0 = \frac{1}{2D}(P_1^2 + P_2^2) + \frac{\beta}{2}(q_1^2 + q_2^2) - \beta(\frac{q_1^3}{3} + q_1 q_2^2), \tag{4.9}$$

and suggests the use of a Melnikov analysis, Guckenheimer and Holmes [1983]. To see this we apply the canonical transformation

$$q_1 = \frac{1}{\sqrt{2D}}(u+v), \qquad q_2 = \frac{1}{\sqrt{2D}}(u-v)$$

to obtain the system

$$\begin{aligned} H = (\frac{1}{2}\dot{u}^2 + \frac{\beta}{2D}u^2 - \frac{2\beta}{3(2D)^{3/2}}u^3) \;&+\; (\frac{1}{2}\dot{v}^2 + \frac{\beta}{2D}v^2 + \frac{2\beta}{3(2D)^{3/2}}v^3) \\ &+\; \frac{\epsilon}{2D}(u-v)^2, \quad \cdot \equiv \frac{d}{dx}. \end{aligned} \tag{4.10}$$

Details of the analysis of (4.10) are to be found in Sleeman [1989a] and follow the lines of a similar problem treated by Holmes [1982].

Briefly, it turns out that the product of a homoclinic orbit $(u, \dot{u})$ and a periodic orbit $(v, \dot{v})$ of the integrable system enjoy very complicated dynamics under the perturbation ϵ. In particular there exists a stochastic layer near the homoclinic manifold containing transverse orbits implying the existence of Smale horseshoes and consequently chaos.

Much remains to be investigated regarding (4.8), not least its biological significance. For instance it is important to investigate the structure of the phase space of (4.8) as the parameters β and τ vary. Thus again in analogy with the case $D = 0$ it is of interest to analyze the dynamics of (4.8) as the parameter $\beta\tau$ varies through the well-known bifurcating sequence leading to period-doubling cascades. Some preliminary numerical work has been performed by Mitchell, Murray and Sleeman [1991] using symplectic schemes.

If we consider the spatial spread of higher period solutions to (4.1) then we are led to consider non-linear "reversible systems" which are non-Hamiltonian perturbations of completely integrable Hamiltonian systems. Such systems are examples of the general theory to be developed in the next section and may well enjoy a property similar to "Arnold diffusion". That is the existence of transition chains of invariant tori carrying spatial chaos from one generation to the next. A motivating example of this is to suppose that (4.1) has solutions that are period-3 in n. (Note: When $D = 0$ period-3 implies chaos.)

In this case we have the system

$$\begin{aligned} u_1 &= \tau D\frac{\partial^2 u_0}{\partial x^2} + u_0 + \tau f(u_0), \\ u_2 &= \tau D\frac{\partial^2 u_1}{\partial x^2} + u_1 + \tau f(u_1), \\ u_0 &= \tau D\frac{\partial^2 u_2}{\partial x^2} + u_2 + \tau f(u_2). \end{aligned} \tag{4.11}$$

Without loss of generality take $D = 1$ and write (4.11) in the equivalent form

$$\begin{aligned} \dot{v}_i &= -\frac{\partial H_0}{\partial u_i} + \frac{1}{\tau}(u_i - u_{i+1}), \\ \dot{u}_i &= \frac{\partial H_0}{\partial v_i}, \quad i = 0, 1, 2, \quad u_3 = u_0, \end{aligned} \tag{4.12}$$

where $H_0 = \sum_{i=0}^{2}[\frac{1}{2}v_i^2 + F(u_i)]$, with

$$F(u) = \int_0^u f(s)ds. \tag{4.13}$$

It is of fundamental importance to recognize that H_0 is simply three copies of the non-linear oscillator

$$H = \frac{1}{2}p^2 + F(q),$$

and, being separable, is completely integrable. Of course, in general, H_0 need not be of such a simple form but could represent the sum of a number of different oscillators such as considered by Holmes and Marsden [1982].

Without being too specific suppose for $i = 1, 2$ we introduce the "action-angle" variables

$$\begin{aligned} I_i &= I_i(u_i, v_i), \\ \theta_i &= \theta_i(u_i, v_i), \end{aligned} \tag{4.14}$$

and write

$$H_0 = \frac{1}{2}p^2 + F(q) + G(I_1) + G(I_2). \tag{4.15}$$

Then the system (4.12) is equivalent to

$$\begin{aligned} \dot{p} &= -\frac{\partial H}{\partial q} - \frac{1}{\tau}(q - u_1(I_1, \theta_1)), \\ \dot{q} &= \frac{\partial H_0}{\partial p}, \\ \dot{I} &= \frac{1}{\tau}g^I(p, q, I_1, I_2, \theta_1, \theta_2), \\ \dot{\theta} &= \Omega(p, q, I_1, I_2) + \frac{1}{\tau}g^\theta(p, q, I_1, I_2, \theta_1, \theta_2), \end{aligned} \tag{4.16}$$

where

$$\begin{aligned} g^I &= (-\frac{\partial I_1}{\partial v_1}(u_2 - u_1), -\frac{\partial I_2}{\partial v_2}(q - u_2))^T, \\ g^\theta &= (-\frac{\partial \theta_1}{\partial v_1}(u_2 - u_1), -\frac{\partial \theta_2}{\partial v_2}(q - u_2))^T, \\ \Omega &= (\frac{\partial G}{\partial I_1}, \frac{\partial G}{\partial I_2}). \end{aligned}$$

The system (4.16) is typical of the general systems we wish to discuss in the next section.

5 The General Theory

Consider the dynamical system

$$\begin{aligned} \dot{x} &= JD_xH(x, I) + \epsilon g^x(x, I, \theta, \mu; \epsilon), \\ \dot{I} &= \epsilon g^I(x, I, \theta, \mu; \epsilon), \\ \dot{\theta} &= \Omega(x, I) + \epsilon g^\theta(x, I, \theta, \mu; \epsilon), \end{aligned} \tag{5.1}$$

where $(x, I, \theta) \in \mathbb{R}^{2n} \times \mathbb{R}^m \times T^l$ and $0 < \epsilon << 1$. In this formulation $\mu \in \mathbb{R}^p$ is a vector of parameters and J is the $2n \times 2n$ symplectic matrix given by

$$J = \begin{pmatrix} 0 & Id \\ -Id & 0 \end{pmatrix},$$

where Id is the $n \times n$ identity matrix and 0 the $n \times n$ zero matrix. In application to the genetic models described in Section 4, $m = l$ and $I \in \mathbb{R}^m$ are action variables with associated angles $\theta \in \mathbb{R}^m$. The system (5.1) is formally comparable with systems of type I treated by Wiggins [1988] and indeed there are considerable similarities. Nevertheless there are important differences which will become apparent as we proceed.

We make the following hypotheses

H1: Let $V \subset \mathbb{R}^{2n} \times \mathbb{R}^m$ and $W \subset \mathbb{R}^p \times \mathbb{R}$ be open sets; then the functions

$$\begin{aligned} JD_xH &: V \to \mathbb{R}^{2n}, \\ g^x &: V \times T^l \times W \to \mathbb{R}^{2n}, \\ g^I &: V \times T^l \times W \to \mathbb{R}^m, \\ \Omega &: V \to \mathbb{R}^l, \\ g^\theta &: V \times T^l \times W \to \mathbb{R}^l, \end{aligned}$$

are defined to be sufficiently differentiable on their respective domains of definition. Furthermore the defining domains of g^x, g^I, g^θ mean that they are each 2π-periodic in each component of their l-dimensional θ arguments.

H2: $H = H(x, I)$ is a scalar valued function which can be thought of as an m-parameter family of Hamiltonians.

We now briefly outline the geometry of the unperturbed phase space where $\epsilon = 0$. That is consider the system

$$\begin{aligned} \dot{x} &= JD_xH(x, I), \\ \dot{I} &= 0, \\ \dot{\theta} &= \Omega(x, I) \end{aligned} \tag{5.2}$$

together with the additional hypotheses.
H3: There exists an open set $U \subset \mathbb{R}^m$ such that for each $I \subset U$ the system

$$\dot{x} = JD_xH(x, I) \tag{5.3}$$

is a completely integrable Hamiltonian system.
H4: For every $I \subset U$ (5.2) possesses a hyperbolic fixed point which may vary smoothly with I and has an n-dimensional homoclinic manifold connecting the fixed point to itself. We shall assume that trajectories along the homoclinic orbit can be represented in the form $x^I(t, \alpha)$ where $t \in \mathbb{R}$, $\alpha \in \mathbb{R}^{n-1}$.

For a fixed $I = I_0 \subset U$ the system (5.2) defines a vector field on $\mathbb{R}^{2n}$. Let x_0 be a hyperbolic fixed point of (5.2) and denote the n-dimensional stable and unstable manifolds of x_0 by $W^s_{I_0}(x_0)$ and $W^u_{I_0}(x_0)$ respectively.

Proposition 5.1 *Suppose $W^s_{I_0}(x_0)$ and $W^u_{I_0}(x_0)$ intersect. Then $W^s_{I_0}(x_0)$ and $W^u_{I_0}(x_0)$ coincide along the n-dimensional components of $W^s_{I_0}(x_0) - \{x_0\}$ and $W^u_{I_0}(x_0) - \{x_0\}$ which contains $W^s_{I_0}(x_0) \cap W^u_{I_0}(x_0) - \{x_0\}$. Hence there exists an n-dimensional manifold of orbits homoclinic to x_0.*

Following Wiggins [1988] we can now describe the phase space of the unperturbed problem in the full (x, I, θ) phase space. To do this we consider the set of points M in $\mathbb{R}^{2n} \times \mathbb{R}^m \times T^l$, namely

$$M = \{(x, I, \theta) \in \mathbb{R}^{2n} \times \mathbb{R}^m \times T^l \mid x = \gamma(I) \quad \text{solves} \quad D_xH(\gamma(I), I) = 0 \textit{ subject to } \det[D_x^2 H(\gamma(I), I)] \neq 0, \forall I \in U, \theta \in T^l\}.$$

Proposition 5.2 *M is an $m + l$ dimensional normally hyperbolic invariant manifold of (5.2). Furthermore M has $n + m + l$ dimensional stable and unstable manifolds $W^s(M)$, $W^u(M)$ which intersect in the $n+m+l$ dimensional homoclinic manifold*

$$\Gamma = \{(x^I(-t_0, \alpha), I, \theta_0) \in \mathbb{R}^{2n} \times \mathbb{R}^m \times T^l \mid (t_0, \alpha, I, \theta_0) \in \mathbb{R}^1 \times \mathbb{R}^{n-1} \times U \times T^l\}.$$

We now turn to the geometric structure of the perturbed phase space ($\epsilon \neq 0$). Under the perturbation the manifold M persists locally as follows:

Proposition 5.3 *There exists $\epsilon_0 > 0$ such that for $0 < \epsilon \leq \epsilon_0$ the perturbed system (5.1) possesses an $m + l$ dimensional normally hyperbolic locally invariant manifold*

$$M_\epsilon = \{(x, I, \theta) \in \mathbb{R}^{2n} \times \mathbb{R}^m \times T^l \mid x = \overline{\gamma}(I, \theta, \epsilon) = \gamma(I) + O(\epsilon), I \subset \overline{U} \subset U \subset \mathbb{R}^m \theta \in T^l\},$$

where $\overline{U} \subset U$ is a compact connected m dimensional set. Moreover M_ϵ has local stable and unstable manifolds $W^s_{loc}(M)$ and $W^u_{loc}(M)$ respectively.

We now want to consider whether there are any recurrent motions on M_ϵ. As pointed out in Wiggins [1988] it is possible there are no recurrent motions on M_ϵ, i.e. all orbits eventually leave M_ϵ by crossing its boundary. In this case the dynamical consequences of the intersection of $W^s(M_\epsilon)$ and $W^u(M_\epsilon)$ are far from clear.

To determine the existence of such motions M_ϵ we could proceed in the usual way and consider the perturbed vector field restricted to M_ϵ. That we consider

$$\begin{aligned} \dot{I} &= \epsilon g^I(\gamma(I), I, \theta; \mu; 0) + O(\epsilon^2), \\ \dot{\theta} &= \Omega(\gamma(I), I) + O(\epsilon), \qquad (I,\theta) \subset \overline{U} \times T^l, \end{aligned} \tag{5.4}$$

and average over the angle variables to get

$$\dot{I} = \epsilon G(I), \tag{5.5}$$

where

$$G(I) = \frac{1}{(2\pi)^l} \int_0^{2\pi} \cdots \int_0^{2\pi} g^I(\gamma(I), I\theta, \mu; 0) d\theta_1 \dots d\theta_l.$$

The persistence of recurrent motions is then assured by the results of Arnold and Avez [1968] or Grebenikov and Ryakov [1983]. Unfortunately in our application $G(I) = 0$ and so another approach is required. If our system were Hamiltonian then the failure of averaging is compensated by invoking the Kolmogorov-Arnold-Moser theorem (provided certain non resonance conditions among the frequencies Ω hold) and again the persistence of recurrent motions is assured, (c.f. Wiggins [1988], p. 386). Since our system is not Hamiltonian we must resort to another device. On returning to the flow (5.4) restricted to M_ϵ we assume it to be reversible in the sense that

$$g^I(\gamma(I), I, -\theta, \mu; 0) = -g^I(\gamma(I), I, \theta, \mu; 0). \tag{5.6}$$

For $\epsilon = 0$ the system admits

$$\begin{aligned} \dot{I} &= 0, \\ \dot{\theta} &= \Omega(\gamma(I), I), \end{aligned} \tag{5.7}$$

where $I = C$ is an invariant torus with flow

$$\dot{\theta} = \Omega(\gamma(C), C),$$

with frequencies $\omega_k = \Omega_k(\gamma(C), C)$. We assume the non-resonance condition

$$\left| \sum_{k=1}^{l} j_k \omega_k \right| \geq \gamma |j|^{-r} \tag{5.8}$$

and then

Theorem 5.1 *(Moser [1973], p. 49). Under the condition (5.8), if*

$$\det \left(\frac{\partial \Omega_k}{\partial I_l} \right) \neq 0 \qquad \text{at} \qquad I = C$$

there exists an invariant torus

$$\theta = \omega_k t + U(\omega_k t, \epsilon) \tag{5.9}$$

$$I = C + V(\omega_k t, \epsilon) \tag{5.10}$$

where U and V are real analytic of period 2π in $\omega_k t$ and vanish for $\epsilon = 0$.

Remarks

1. The motion on these surviving tori is quasi-periodic having l rationally incommensurate frequencies in the sense of (5.8). The invariant tori form a majority in the sense that the Lebesgue measure of the complement of their union is small when ϵ is small.
2. If the non-resonance condition (5.8) does not hold then new techniques are required to reveal some very interesting dynamics yet to be explored. In our motivating example of Section 4 we have resonances but these are detuned under the perturbation and recurrent motions are expected to persist.

Theorem 5.1 is the analogue of the KAM theory for reversible systems.

We are now in a position to analyze the splitting of manifolds under the perturbation ϵ. This analysis, which is rather involved, is developed fully in Sleeman [1996] and requires, as in Holmes and Marsden [1982] or Wiggins [1988], the construction of a vector of Melnikov integrals. Specifically these are written as

$$M_l^{\overline{I}}(\theta_0, \alpha; \mu) = \int_{-\infty}^{\infty} [< D_x K_i, g^x > + < D_I K_i, g^I >](q_0^{\overline{I}}(t), \mu, 0)dt$$
$$- < D_I K_i(\gamma(\overline{I}), \overline{I}), \quad \int_{-\infty}^{\infty} g^I(q_0^{\overline{I}}(t), \mu; 0)dt >, \quad i = 1, \dots n, \tag{5.11}$$

where

$$q_0^{\overline{I}}(t) = \left(x^{\overline{I}}(t, \alpha), \overline{I}, \int^t \Omega(x^{\overline{I}}(s, \alpha)ds + \theta_0 \right), \qquad \overline{I} \in \overline{U} \subset U \subset \mathbb{R}^m,$$

and $K_1 = H$, K_2, ... , K_m are the integrals of the system (5.3).

In addition we have

$$M_i^{\overline{I}}(\theta_0, \alpha; \mu) = \int_{-\infty}^{\infty} (g^{\overline{I}})_{i-n}(q_0^{\overline{I}}(t - t_0), \mu; 0)dt, \qquad i = n + 1, \dots n + m. \tag{5.12}$$

We can now state our fundamental results.

Theorem 5.2 *Suppose there exists $I = \overline{I} \in \overline{U} \subset \mathbb{R}^m$ such that $\tau_\epsilon(\overline{I})$, a surviving torus on M_ϵ, is a KAM torus. Let $(\theta_0, \alpha, \mu) = (\overline{\theta}_0, \overline{\alpha}, \overline{\mu}) \in T^m \times \mathbb{R}^{n-1} \times \mathbb{R}^p$ be a point with $m + n + p - 1 \geq m + n$ such that*
1. *$M^{\overline{I}}(\overline{\theta}_0, \overline{\alpha}, \overline{\mu}) = 0$,*
2. *$DM^{\overline{I}}(\overline{\theta}_0, \overline{\alpha}, \overline{\mu})$ is of rank $m + n$.*

Then $W^s(\tau_\epsilon(\overline{I}))$ and $W^u(\tau_\epsilon(\overline{I}))$ intersect near $(\overline{\theta}_0, \overline{\alpha}, \overline{\mu})$.

To guarantee transverse intersections we have

Theorem 5.3 *Suppose Theorem 5.2 holds at the point*

$$(\theta_0, \alpha, \mu) = (\overline{\theta}_0, \overline{\alpha}, \overline{\mu}) \in T^m \times \mathbb{R}^{n-1} \times \mathbb{R}^p$$

and that

$$D_{(\theta_0, \alpha)} M^{\overline{I}}(\overline{\theta}_0, \overline{\alpha}, \overline{\mu})$$

is of rank $m + n$. Then for ϵ sufficiently small $W^s(\tau_\epsilon(\overline{I}))$ and $W^u(\tau_\epsilon(\overline{I}))$ intersect transversely near $(\overline{\theta}_0, \overline{\alpha})$.

These results provide the possible underlying mechanism for chaos. In many cases, including the biological examples we have in mind, that the system is chaotic follows from the use of the lambda-lemma as described in Wiggins [1988]. In general the transverse crossing of stable and unstable manifolds is not sufficient to assert the existence of Smale horseshoes as the famous example of Devaney [1978] shows.

As an application of this theory we return to the system (4.11) which when recast in terms of action angle variables has the form (4.16) which has the unperturbed Hamiltonian (4.15). With the choice $f(x_1) = \beta x_1(1-x_1)$ then

$$H_0 = \frac{1}{2}x_2^2 + \beta(\frac{x_1^2}{2} - \frac{x_1^3}{3}) + G(I_1) + G(I_2). \tag{5.13}$$

Thus, in the notation of the general system (5.1), $n = 1$, $m = l = 2$.

Consequently H_0 has the hyperbolic fixed point $x_0 = (1,0)$ and a 1-dimensional homoclinic manifold connecting the fixed point to itself and has 1-dimensional stable and unstable manifolds given by $W^s(1,0)$ and $W^u(1,0)$ respectively. Specifically the homoclinic orbit is given by

$$(x_1, x_2) = \left(1 - \frac{3}{2}sech^2(\frac{1}{2}\sqrt{\beta}x), -\frac{3}{2}\sqrt{\beta}tanh(\frac{1}{2}\sqrt{\beta}x)sech^2(\frac{1}{2}\beta x)\right), \tag{5.14}$$

with energy $h_0 = \frac{\beta}{6}$.

The corresponding periodic orbits contributing to the actions $G(I_i)$, $i = 1,2$ are determined in terms of Jacobian elliptic functions and are similar to those orbits computed in Sleeman [1989a] or Holmes [1982]. Thus in the full $(x_1, x_2, \theta_1, \theta_2, I_1, I_2)$ phase space the unperturbed system has a 4-dimensional normally hyperbolic invariant manifold given by

$$M = \{(x_1, x_2, I_1, I_2, \theta_1, \theta_2) \in \mathbb{R}^2 \times \mathbb{R}^2 \in T^2 \mid (x_1, x_2) = (1,0)\} \tag{5.15}$$

which has the structure of a two parameter family of 2-tori. As in Proposition 5.2 M has 5-dimensional stable and unstable manifolds which coincide along the 5-dimensional homoclinic orbit Γ which can be parameterized by

$$\Gamma = \left\{1 - \frac{3}{2}sech^2(\frac{1}{2}\sqrt{\beta}x) \quad , \quad -\frac{3}{2}\sqrt{\beta}tanh(\frac{1}{2}\sqrt{\beta}x)sech^2(\frac{1}{2}\sqrt{\beta}x), \right. \tag{5.16}$$

$$\left. I_1, I_2, \theta_1^0, \theta_2^0 \in \mathbb{R}^2 \times \mathbb{R}^2 \times T^2 \mid x \in \mathbb{R}\right\}.$$

Note that on M, for each I_1, I_2 there corresponds a torus $\tau(I)$ having 3-dimensional stable and unstable manifolds coinciding along the 3-dimensional homoclinic orbit Γ_I obtained from (5.14) by fixing the I components.

The theory now proceeds, as in the general case, to the construction of the Melnikov vector. In this example they take the simple form

$$M_1^{\overline{I}} = \int_{-\infty}^{\infty} x_2(q_2 - x_1)dt, \tag{5.17}$$

$$M_2^{\overline{I}} = \int_{-\infty}^{\infty} p_2(q_3 - q_2)dt,$$

$$M_3^{\overline{I}} = \int_{-\infty}^{\infty} p_3(x_1 - q_3)dt.$$

(5.18)

These integrals are rather complicated and involve both elliptic functions and hyperbolic functions. They can nevertheless be evaluated using residue theory (details given in Sleeman [1996]). The upshot is that they do possess simple zeros in the sense of Theorems 5.2 and 5.3 and consequently there is the possibility of spatial chaos in the migration of the gene density.

Finally we remark that the theory outlined in this section appears to have numerous applications and implications both in biology and physiology. We hope to explore these applications in future work.

In the next section we turn to the study of travelling waves in the ecological models of Section 3 which, as we shall see, suggest further open problems related to Hamiltonian systems.

6 Travelling Waves in Ecology

Motivated by the work of Weinberger [1978] and more recently by Kot [1992] we consider the existence of travelling waves to the system (3.1), (3.2), (3.4). That is we seek solutions of the form

$$N_{t+1}(x) = N_t(x+c) \tag{6.1}$$

for some wave speed c. Following Weinberger [1978] it is not difficult to prove that our system supports a continuum of travelling waves with minimum speed of propagation given by

$$C^* = \min_{\mu>0}\left\{\frac{1}{\mu}\ln[(1+r)\int_{-\infty}^{\infty} k(s)e^{-\mu s}ds]\right\}.$$

Kot [1992] in his analysis, using a combination of singular perturbation techniques and numerical computations, has shown that the travelling waves (6.1) enjoy some very intriguing properties which deserve further investigation. Here we provide some preliminary results in showing how travelling waves may be studied through Hamiltonian systems.

First of all it is easy to see that the system (3.1), (3.2), (3.4) under the assumed form (6.1) is equivalent to the second-order delay-differential equation

$$N^{''}(x+c) + \alpha^2[f(N(x)) - N(x+c)] = 0. \tag{6.2}$$

If we normalize by setting $z = \frac{x}{c}$, $\epsilon = \frac{1}{\alpha^2}c^2$ then

$$\epsilon N^{''}(z+1) + [f(N(z)) - N(x+1)] = 0. \tag{6.3}$$

Period-1 solutions of (6.3) are equivalent to the study of **travelling fronts**. That is with $N(z) = N(z+1)$ we have

$$\epsilon N^{''} + (f(N) - N) = 0. \tag{6.4}$$

This equation is "Hamiltonian" and can be solved by quadratures in precisely the same manner as (4.5). In his numerical experiments Kot [1992] found the appearance of period-2 cycles to the equation (6.3) as r (c.f. (3.2)) was increased. Motivated by this we look for period-2 solutions of (6.4). That is we seek solutions N_1, N_2 such that

$$\begin{aligned} \epsilon N_2^{''} + (f(N_1) - N_2) &= 0 \\ \epsilon N_1^{''} + (f(N_2) - N_1) &= 0 \end{aligned} \tag{6.5}$$

This system is also Hamiltonian and has the form

$$\begin{aligned} M_1' &= -\frac{\partial H}{\partial N_1}, \qquad M_2' = -\frac{\partial H}{\partial N_2}, \\ N_1' &= \frac{\partial H}{\partial M_1}, \qquad N_2' = \frac{\partial H}{\partial M_2}, \qquad ' \equiv \frac{d}{dz} \end{aligned} \tag{6.6}$$

where

$$H = \frac{1}{\epsilon}M_1M_2 - N_1N_2 + \frac{(1+r)}{2}(N_1^2 + N_2^2) - \frac{r}{3K}(N_1^3 + N_2^3). \tag{6.7}$$

The novel feature of this system is that H is of the form

$$H = KE + PE \tag{6.8}$$

where kinetic energy (KE) is indefinite.

Hamiltonians of this form are relatively new to the literature but have been considered by Hofer and Toland [1984], Chow and Kim [1988], Mielke, Holmes and O'Reilly [1992] and Bridges [1992]. Nevertheless such systems are little understood. We can prove the existence of heteroclinic orbits for (6.7) thus confirming the period-2 cycles discovered by Kot [1992]. However for $k = 1$ and $2 < r < \sqrt{6}$ the system may not be completely integrable in the sense of Ziglin [1982] [1983] a subject of current investigation (Rod and Sleeman [1995]).

Finally we draw attention to travelling waves that may arise in a predator-prey system in which a predator and a prey grow and disperse in synchrony but with different diffusivities. Such models are of the form

$$\begin{aligned} N_{t+1} &= N_t \exp[r(1 - N_t - P_t)] \\ P_{t+1} &= bN_tP_t \end{aligned} \tag{6.9}$$

where N_t is the number of prey and P_t is the number of predators at time t. In the absence of the predator, a high r prey ($r > 2$) exhibits oscillatory or chaotic behaviour.

In the presence of the predator and dispersal we may consider the integro-difference model

$$\begin{aligned} N_{t+1}(x) &= \int_{-\infty}^{\infty} k_1(x-y)N_t(y)\exp\{r[1 - N_t(y) - P_t(y)]\}dy, \\ P_{t+1}(x) &= \int_{-\infty}^{\infty} k_2(x-y)bN_t(y)P_t(y)dy. \end{aligned} \tag{6.10}$$

In studies of diffusive instability Kot [1989] has shown that over dispersal of the predator relative to the prey can lead to a cascade of diffusion driven period-doubling bifurcations.

The system (6.10) has much in common with the cellular automata system described in Section 2. However here we believe considerable progress can be made in the understanding of predator-prey interactions using the dynamical systems approach advocated in this paper. This is currently under investigation.

Postscript

The study of spatio-temporal chaos observed in biology through the use of ideas drawn from Hamiltonian systems and reversible systems is new. It offers numerous and challenging problems for the analyst and, we believe, has important implications for the life and medical sciences.

Acknowledgement

I would like to thank Gyorgy Haller, Tom Bridges and Richard Cushman for useful discussions on the presence of Smale horseshoes in our general theory.

References

Arnold, V. I., and Avez, A. [1968], *Ergodic problems of classical mechanics,* W. A. Benjamin, New York.

Bridges, T. J. [1992], *Hamiltonian bifurcations of the spatial structure for coupled non linear Schrodinger equations,* Physica D, **57**, 375–394.

Chow, S., and Kim, Y. I. [1988], *Bifurcation of periodic orbits for non-positive definite Hamiltonian systems,* App. Anal., **31**, 163–199.

Comins, H. N., Hassell, M. P., and May, R. M. [1992], *The spatial dynamics of host-parasitoid systems,* J. Animal Ecology, **61**, 735–748.

Devaney, R. [1978], *Transversal homoclinic orbits in an integrable system,* Amer. J. Math., **100**, 631–642.

Grebenikov, E. A., and Ryakov, Yu. A. [1983], *Constructive methods in the analysis of non linear systems,* Mir, Moscow.

Guckenheimer, J., and Holmes, P. J. [1983], *Nonlinear oscillations, dynamical systems and bifurcations of vector fields,* Springer Verlag, New York, Heidelberg, Berlin.

Hassell, M. P., Comins, H. N., and May, R. M. [1991], *Spatial structure and chaos in insect population dynamics,* Nature, **353**, 255–258.

Hofer, H., and Toland, J. [1984], *Homoclinic, heteroclinic and periodic orbits for a class of indefinite Hamiltonian systems,* Math. Ann., **268**, 387–403.

Holmes, P. J. [1982], *Proof of non integrability for the Henon-Heiles Hamiltonian near an exceptional integrable case,* Physica, **5D**, 335–347.

Holmes, P. J., and Marsden, J. E. [1982], *Melnikov's method and Arnold diffusion for perturbations of integrable Hamiltonian systems,* J. Math. Phys., **23**, 669–675.

Kot, M. [1989], *Diffusion-drive period-doubling bifurcations,* Biosystems, **22**, 279–287.

Kot, M. [1992], *Discrete-time traveling waves: Ecological examples,* J. Math. Biol., **30**, 413–436.

May, R. M. [1976], *Simple mathematical models with very complicated dynamics,* Nature, **26**, 459–467.

Mielke, A., Holmes, P. J., and O'Reilly, O. [1992], *Cascades of homoclinic orbits to, and chaos near, a Hamiltonian saddle center,* J. Dynamics and Diff. Equations, **4**, 95–126.

Mitchell, A. R., Murray, B. A., and Sleeman, B. D. [1991], *Numerical solution of Hamiltonian systems in reaction-diffusion by symplectic difference schemes,* J. Comp. Phys., **95**, 339–358.

Moser, J. [1973], *Stable and random motions in dynamical systems,* Princeton University Press, Princeton.

Rod, D.L., and Sleeman, B.D. [1995], *Complexity in spatio-temporal dynamics,* Proc. Roy. Soc. Ed. **125A**, to appear.

Sleeman, B. D. [1989a], *Complexity in biological systems and Hamiltonian dynamics,* Proc. Roy. Soc. A, **425**, 17–47.

Sleeman, B. D. [1989b], *Spatio-temporal complexity in biological systems,* in Nonlinear Waves in Active Media (ed. J. Engelbrecht), Research Reports in Physics, Springer-Verlag, New York, Heidelberg, Berlin, 192–200.

Sleeman, B. D. [1996], *Complexity of reversible dynamical systems and integrability,* in Proceedings of the International Conference on Differential Equations and Control Theory", World Scientific, to appear.

Weinberger, H. F. [1978], *Asymptotic behaviour of a model of population genetics,* in Non-linear Partial Differential Equations and Applications (ed. J. Chadam), Lect. Notes. in Math., Springer-Verlag, New York, Heidelberg, Berlin, **648**, 47–98.

Wiggens, S. [1988], *Global bifurcations and chaos,* Springer-Verlag, New York, Heidelberg, Berlin.

Yee, H. C, Sweby, P. K., and Griffiths, D. F. [1991], *Dynamical approach study of spurious steady-state numerical solutions of non-linear differential equations,* The Dynamics of Time Discretization and its Implications for Algorithm Development in Computational Fluid Dynamics, J. Comp. Phys, **97**, 249–310.

Ziglin, S. L. [1982], *Branching of solutions and non-existence of first integrals in Hamiltonian Mechanics I,* Functional Anal. Appl., **16**, 6–17.

Ziglin, S. L. [1983], *Branching of solutions and non-existence of first integrals in Hamiltonian Mechanics II,* Functional Anal. Appl., **17**.

Fields Institute Communications
Volume 8, 1996

Degenerate Hamiltonian Hopf Bifurcations

J.C. van der Meer
Eindhoven University of Technology
P.O.Box 513, 5600 MB Eindhoven
The Netherlands

1 Introduction

In Van der Meer [1985] the (nondegenerate) Hamiltonian Hopf bifurcation was studied. On $\mathbb{R}^4$ with canonical symplectic matrix J the nondegenerate Hamiltonian Hopf bifurcation is described by the Hamiltonian system $\dot{z} = J\nabla H(z)$ with Hamiltonian

$$H(x,y) = (x_1y_2 - x_2y_1) + \frac{1}{2}(x_1^2 + x_2^2) + \frac{\nu}{2}(y_1^2 + y_2^2) + a(y_1^2 + y_2^2)^2 + h.o.t. \quad (1)$$

At $\nu = 0$ we have a collision of eigenvalues of opposite signature on the imaginary axis. For $\nu > 0$ the eigenvalues lie on the imaginary axis, while for $\nu < 0$ the eigenvalues lie in the complex plane. When a is nonzero we call the bifurcation nondegenerate. In this case the above Hamiltonian without higher order terms can be taken as a normal form for the bifurcation. See Van der Meer [1985] for more details. A bifurcation like this occcurs for instance in the circular restricted three body problem at Routh's critical mass ratio (Van der Meer [1985], ch.6), in the Lagrange top (Cushman and Van der Meer [1990],Van der Meer [1990]), and in the double spherical pendulum (Marsden [1992]).

When a is zero we speak of a degenerate Hamiltonian Hopf bifurcation. This was first studied in Bridges [1990], Bridges [1991], motivated by problems from fluid mechanics. He used quite different methods than in Van der Meer [1985]. In the degenerate case the normal form for the bifurcation will contain higher order terms and the bifurcation will be of higher codimension. In the degenerate case as considered in this paper the next power of $(y_1^2 + y_2^2)$ in the normal form, i.e. $(y_1^2 + y_2^2)^3$, is supposed to have nonzero coefficient.

1991 *Mathematics Subject Classification.* 58F05, 58F14, 58F22.

Key words and phrases. Hamiltonian system, bifurcation theory, singularity theory, Hamiltonian Hopf bifurcation, periodic solutions.

The research in this paper was supported in part by the E.E.C. Science project on "Bifurcation Theory and its Applications".

In this paper we deal with the problem of the degenerate Hamiltonian Hopf bifurcation along the lines of Van der Meer [1985]. Basically the approach consists of four steps:

1. Normal form theory for (Hamiltonian) vector fields.
2. Weinstein-Moser reduction.
3. Singularity theory.
4. Analysis of the singularity (bifurcation) theoretic normal form.

The first two steps apply to a broad class of vector fields, also non-Hamiltonian ones. Because these are the same as in the nondegenerate case we will focus on the last two steps and only review the first two steps in section 2.

It turns out that it is sufficient to do the singularity theory in terms of topological equivalence. However, one first does the formal calculations as in the C^∞ context. From this one can derive the topological normal form using results of Damon [1988a], Damon [1988b]. We find the normal form below, which has topological codimension two, λ_1, and λ_2 being the unfolding parameters.

$$\begin{aligned} H(x,y) =&(x_1y_2 - x_2y_1) + \frac{1}{2}(x_1^2+x_2^2) + \frac{\lambda_1}{2}(y_1^2+y_2^2) + \\ &\frac{\lambda_2}{8}(y_1^2+y_2^2)^2 + \frac{a}{2}(y_1^2+y_2^2)(x_1y_2-x_2y_1) + \\ &\frac{b}{24}(y_1^2+y_2^2)^3 \end{aligned} \tag{2}$$

We have the nondegeneracy conditions $b \neq 0, a^2+b \neq 0, 3a^2-b \neq 0$. Actually the cases $a=0, a \neq 0$ are topologically nonequivalent, the case $a=0$ having codimension three.

This paper outlines results concerning a complete program for analyzing Hamiltonian Hopf bifurcations. Complete details and much more on this subject will appear in a forthcoming joint publication with Jan Sanders and André Vanderbauwhede.

2 Normalization and reduction to an integrable system

In this section we will make some remarks about the first two steps mentioned above in order to make clear what our starting point for the remaining two steps will be.

Consider a parameter dependent Hamiltonian system on $\mathbb{R}^4$ with the standard symplectic form. We suppose the system to have a stationary point at the origin for each value of the parameter, i.e. for the Hamiltonian function $H(z;\delta)$ we have $d_zH(0;\delta)=0$, δ being the parameter. We write the Taylor series expansion of H at 0 as

$$H(z;\delta) = H_2(z;\delta) + \cdots + H_k(z;\delta) + h.o.t., \tag{3}$$

where H_k stand for the homogeneous terms of degree k. We speak of a Hamiltonian Hopf bifurcation at $\delta=0$ if

(i) $H_2(z;0)$ has a nontrivial Jordan-Chevalley decomposition in a semisimple part $S(z)$ corresponding to a linear system with a double pair of purely imaginary eigenvalues $\pm\alpha i$ and a nonzero nilpotent part $N(z)$.

(ii) For δ passing through zero we have a collision of purely imaginary eigenvalues of opposite signature at αi and $-\alpha i$ after which the eigenvalues move off the axis into the complex plane.

Note that at a Hamiltonian Hopf bifurcation the linear stability type at the stationary point changes from stable to unstable (or vice versa). Without loss of generality we may suppose α to be 1. Furthermore we suppose $H_2(z;0)$ to be in linear normal form (Williamson [1936],Burgoyne and Cushman [1974]), i.e.

$$H_2(x,y;0) = (x_1y_2 - x_2y_1) + \frac{1}{2}(x_1^2 + x_2^2), \tag{4}$$

with

$$\begin{aligned} S(x,y) &= x_1y_2 - x_2y_1, \\ N(x,y) &= \frac{1}{2}(x_1^2 + x_2^2). \end{aligned} \tag{5}$$

Note that a minus sign in front of N gives a non-equivalent normal form. We have chosen the normal form with the plus sign for reasons of presentability. The normal form with a minus sign can be dealt with in an analogous way.

Once we know the quadratic part $H_2(z;0)$ we may put our Hamiltonian function $H(z;0)$ into normal form (Cushman et al. [1974],Van der Meer [1985], Cushman and Sanders [1987]). For this we need to know an $sl(2,\mathbb{R})$ embedding of N. In our case we may choose as $sl(2,\mathbb{R})$ generators

$$N(x,y), M(x,y) = \frac{1}{2}(y_1^2 + y_2^2), T(x,y) = x_1y_1 + x_2y_2. \tag{6}$$

Let $\{,\}$ denote the standard Poisson bracket. Then the bracket relations are given by

$$\{M,N\} = T, \{M,T\} = 2M, \{N,T\} = -2N. \tag{7}$$

Furthermore we have

$$\{S,N\} = \{S,M\} = \{S,T\} = 0. \tag{8}$$

Let $\mathcal{S}$ denote the one parameter group generated by the Hamiltonian vector field corresponding to S, the $\mathcal{S}$-action is an S^1-action. The functions S, N, M, and T generate the space of polynomials invariant under the $\mathcal{S}$-action. Thus according to a theorem of Schwarz [1975] any smooth function in ker(ad(S)), i.e. $\mathcal{S}$-invariant function, is a smooth function of S, N, M, and T. Here ad(S) is the operator given by $\{S,\cdot\}$.

Definition 2.1

(i) *$H_k(z;0)$ is in normal form with respect to $H_2(z;0)$ if $H_k(z;0) \in ker(ad(S)) \cap ker(ad(M))$.*

(ii) *$H(z;0)$ is in normal form up to order k with respect to $H_2(z;0)$ if H_m, $2 < m < k+1$, is in normal form with respect to $H_2(z;0)$.*

Consequently the normal form of H_k must be a polynomial in S and M and the normal form for $H(z;0)$ up to order $2k$ can be written as

$$\bar{H}(z;0) = S + N + \bar{H}_2(M,S) + \cdots + \bar{H}_k(M,S) + h.o.t., \tag{9}$$

where H_k stand for the homogeneous terms of degree k in M and S. Normal form theory guarantees the existence of a symplectic normalizing transformation. For actual computations constructive algorithms which find the normal form and the normalizing transformation have been developed which can be implemented using symbolic computation packages (see Cushman and Sanders [1988] for instance).

For the parameter dependent system $H(z;\delta)$, even if $H_2(z;\delta)$ is a nontrivial deformation of $H_2(z;0)$, we may proceed along the same lines and choose our normalized terms to be in $\ker(\mathrm{ad}(S)) \cap \ker(\mathrm{ad}(M))$ up to order $2k$ provided δ is sufficiently small.

Once we have normalized the system we want to know where to truncate this normal form in order to obtain a true description of the bifurcation of periodic orbits. For this we want to make use of singularity theory. However, we first have to establish a connection between the system in normal form and an integrable system which agrees with the normalized system up to the order to which it is put in normal form. This is the key point in the next theorem.

Consider a Hamiltonian system in normal form with Hamiltonian $\bar{H}$ like in (9). The Hamiltonian vector field X_S corresponding to S has only periodic solutions. The periodic solutions of $X_{\bar{H}}$ we are interested in are those close to orbits of X_S with almost the same period. Let $\Sigma_{\bar{H}}$ denote the set of these orbits.

Theorem 2.1 *Consider a Hamiltonian system in normal form up to order $2k$ with Hamiltonian $\bar{H}$ like in (9). There exists a Hamiltonian system with Hamiltonian $\hat{H}$ such that*

(i) $\bar{H} - \hat{H} = O(|z|^{2k+1})$,

(ii) $\{\hat{H}, S\} = 0$,

(iii) $\Sigma_{\bar{H}}$ *and* $\Sigma_{\hat{H}}$ *are* C^{2k} *diffeomorphic.*

A proof of this theorem can be found in Van der Meer [1985], Duistermaat [1984] or Vanderbauwhede and Van der Meer [1995]. Note that the correspondence between the two systems is only on the set of periodic solutions (up to diffeomorphism). The two systems are certainly not equivalent by symplectic maps. However, they can be seen as perturbations of one another.

Obviously $\Sigma_{\hat{H}}$ consists of orbits which lie along the X_S trajectories but have slightly different period. $\Sigma_{\hat{H}}$ is given by the bifurcation equation $d\hat{H} = \tau dS$, where τ is the reciprocal of the period of the $X_{\hat{H}}$ orbit. Thus $\Sigma_{\hat{H}}$ coresponds precisely to the critical set of the energy-momentum mapping

$$\hat{H} \times S : \mathbb{R}^4 \to \mathbb{R}^2; (x, y) \to (\hat{H}(x, y), S(x, y)). \tag{10}$$

In the following we will study the singularity of this map. The eventual bifurcation diagram corresponds to the set of critical values of this map.

3 Singularity theoretic setting

In the following sections we will study the singularities of (10). Unfolding the determining jet provides us with a singularity theoretic normal form for the energy momentum map and the bifurcation set. Interpreting the singularity theoretic normal form for (10) again as an energy momentum mapping for a Hamiltonian system we obtain an unfolding of a finite jet of our Hamiltonian in normal form (9). Comparing this with the corresponding jet of the original parameter dependent normal form gives us the bifurcation set of the original parameter dependent system. References for the singularity theory used in the sequel are Bierstone, [1980], Golubitsky and Guillemin [1973], Martinet [1982], Poènaru [1976], Roberts [1986].

Let $Diff_0(\mathbb{R}^4)_S$ denote the origin preserving $\mathcal{S}$-equivariant diffeomorphisms from $\mathbb{R}^4$ to itself, where $\mathcal{S}$ is the S^1-action given by the flow of X_S. Furthermore let $Diff_0(\mathbb{R}^2)$ denote the origin preserving diffeomorphisms from $\mathbb{R}^2$ to itself. Consider the group action of $\mathcal{A}_S = Diff_0(\mathbb{R}^4)_S \times Diff_0(\mathbb{R}^2)$ on the space of C^∞ mappings from $\mathbb{R}^4$ to $\mathbb{R}^2$ given by $(\phi, \psi) \cdot f = \psi \circ f \circ \phi^{-1}$, with $\phi \in Diff_0(\mathbb{R}^4)_S$, $\psi \in Diff_0(\mathbb{R}^2)$, $f \in C^\infty(\mathbb{R}^4, \mathbb{R}^2)$. For $f \in C^\infty(\mathbb{R}^4, \mathbb{R}^2)$ let $T\mathcal{A}_S(f)$ denote the tangent space at f to the $\mathcal{A}_S$ orbit through f. Two maps $f, g \in C^\infty(\mathbb{R}^4, \mathbb{R}^2)$ are called **equivalent** if f and g are in the same $\mathcal{A}_S$ orbit. The codimension of $T\mathcal{A}_S(f)$ in $C^\infty(\mathbb{R}^4, \mathbb{R}^2)$ is called the **codimension** of f. A map f of codimension zero is called **stable**, stable maps are equivalent to all nearby maps (in the appropriate topology). An r-parameter **unfolding** of f_0 is an $f \in C^\infty(\mathbb{R}^4 \times \mathbb{R}^r, \mathbb{R}^2 \times \mathbb{R}^r)$, such that, $f(x,u) = (\tilde{f}(x,u), u)$ and $f_0(x) = \tilde{f}(x,0)$. Often $\tilde{f}$ is called the unfolding of f_0. The following theorem characterizes **universal** unfoldings.

Theorem 3.1 *Let f be an r-parameter unfolding of f_0. f is a universal unfolding if and only if*

$$T\mathcal{A}_S(f_0) + \mathbb{R}\{\frac{\partial \tilde{f}}{\partial u_1}(x,0), \ldots, \frac{\partial \tilde{f}}{\partial u_r}(x,0)\} = C^\infty(\mathbb{R}^4, \mathbb{R}^2),$$

where $\mathbb{R}\{\frac{\partial \tilde{f}}{\partial u_1}(x,0), \ldots, \frac{\partial \tilde{f}}{\partial u_r}(x,0)\}$ is the $\mathbb{R}$ module generated by the $\frac{\partial \tilde{f}}{\partial u_i}(x,0)$, $i = 1, \ldots, r$.

A universal unfolding is called **minimal** if the number of parameters is minimal. If a map is of finite codimension r, it has an r-parameter minimal universal unfolding. Thus to establish the minimal universal unfolding of a map we have to determine a minimal number of generators for the complement of its tangent space with respect to the $\mathcal{A}_S$ action.

4 Tangent space computations

Let $C^\infty(\mathbb{R}^4, \mathbb{R}^2)^{\mathcal{S}}$denote the space of $\mathcal{S}$-invariant maps in $C^\infty(\mathbb{R}^4, \mathbb{R}^2)$. Consider a map $f \in C^\infty(\mathbb{R}^4, \mathbb{R}^2)^{\mathcal{S}}$, $f(x,y) = (f_1(x,y), f_2(x,y))$. The tangent space $T\mathcal{A}_S(f)$ consists of two terms (cf. Roberts [1986]).

The first term is the tangent space due to the right action of $Diff_0(\mathbb{R}^4)_S$, it is the Jacobian module $\mathcal{J}(f)^{\mathcal{S}}$, i.e. the space of vector fields $df \circ \xi$, with ξ a $\mathcal{S}$-equivariant origin preserving vector field on $\mathbb{R}^4$. $\mathcal{J}(f)^{\mathcal{S}}$ can be considered as a $C^\infty(\mathbb{R}^4)^{\mathcal{S}}$-module.

The second term is the tangent space due to the left action of $Diff_0(\mathbb{R}^2)$. It is the space of vector fields $\eta \circ f$, where η is an origin preserving vector field on $\mathbb{R}^2$. Let $\mathcal{F}(f_1, f_2)$ denote the C^∞ functions in the component functions of f. Then the second term of the tangent space is $\mathcal{M}(f_1, f_2) \times \mathcal{M}(f_1, f_2)$, where $\mathcal{M}(f_1, f_2)$ is the maximal ideal in $\mathcal{F}(f_1, f_2)$. $\mathcal{M}(f_1, f_2) \times \mathcal{M}(f_1, f_2)$ can be considered as a $\mathcal{F}(f_1, f_2)$-module.

We want to study the tangent space to an energy-momentum mapping $\bar{H} \times S$. However, we do not want to compute this map up to infinite order. Therefore we guess a determining part G of $\bar{H}$, and compute the tangent space to $G \times S$. The computations will show whether $\bar{H} \times S$ and $G \times S$ are equivalent and whether we can improve on our choice of G. Take

$$G = N + aSM + \frac{b}{3}M^3, \tag{11}$$

with S, N and, M as in (6). We want to determine the tangent space

$$T\mathcal{A}_{\mathcal{S}}(G \times S) = \mathcal{J}(G \times S)^{\mathcal{S}} + \mathcal{M}(G,S) \times \mathcal{M}(G,S). \tag{12}$$

In order to determine the Jacobian module $\mathcal{J}(G \times S)^{\mathcal{S}}$ we need the following

Proposition 4.1 *A general $\mathcal{S}$-equivariant vector field is given by*

$$\begin{aligned}
&F_1(N,S,T)\begin{pmatrix} x_1 \\ x_2 \\ 0 \\ 0 \end{pmatrix} + F_2(N,S,T)\begin{pmatrix} -x_2 \\ x_1 \\ 0 \\ 0 \end{pmatrix} + \\
&F_3(M,S,T)\begin{pmatrix} y_1 \\ y_2 \\ 0 \\ 0 \end{pmatrix} + F_4(M,S,T)\begin{pmatrix} -y_2 \\ y_1 \\ 0 \\ 0 \end{pmatrix} + \\
&F_5(N,S,T)\begin{pmatrix} 0 \\ 0 \\ x_1 \\ x_2 \end{pmatrix} + F_6(N,S,T)\begin{pmatrix} 0 \\ 0 \\ -x_2 \\ x_1 \end{pmatrix} + \\
&F_7(M,S,T)\begin{pmatrix} 0 \\ 0 \\ y_1 \\ y_2 \end{pmatrix} + F_8(M,S,T)\begin{pmatrix} 0 \\ 0 \\ -y_2 \\ y_1 \end{pmatrix}.
\end{aligned} \tag{13}$$

Actually the above is the Stanley decomposition for an $\mathcal{S}$-equivariant vector field. More on Stanley decompositions can be found in Cushman and Sanders [1990].

Write (13) as $\sum_{i=1}^{8} F_i v_i$. Furthermore let

$$V_i = L_{v_i}\begin{pmatrix} G \\ S \end{pmatrix}, \tag{14}$$

where L_{v_i} is the Lie derivative with respect to the vector field v_i. Then the V_i generate $\mathcal{J}(f)^{\mathcal{S}}$ as a $C^\infty(I\!R^4)^{\mathcal{S}}$-module. In the following we will write $\mathcal{F}(N,M,S,T)$ instead of $C^\infty(I\!R^4)^{\mathcal{S}}$. Straightforward computations give

$$\begin{aligned}
V_1 &= \begin{pmatrix} 2N + aSM \\ S \end{pmatrix}, & V_2 &= -\begin{pmatrix} aMT \\ T \end{pmatrix}, \\
V_3 &= \begin{pmatrix} T \\ 0 \end{pmatrix}, & V_4 &= -\begin{pmatrix} S + 2aM^2 \\ 2M \end{pmatrix}, \\
V_5 &= \begin{pmatrix} aST + bM^2T \\ 0 \end{pmatrix}, & V_6 &= \begin{pmatrix} aS^2 + bSM^2 + 2aNM \\ 2N \end{pmatrix}, \\
V_7 &= \begin{pmatrix} 3aSM + 2bM^3 \\ S \end{pmatrix}, & V_8 &= \begin{pmatrix} aMT \\ T \end{pmatrix}.
\end{aligned} \tag{15}$$

Let $\mathcal{I}$ be the $\mathcal{F}(N, M, S, T)$-module generated by V_2 and $V3$. Clearly $\mathcal{I}$ is a submodule of $\mathcal{F}(N, M, S, T) \times \mathcal{F}(N, M, S, T)$, but also of $\mathcal{J}(G \times S)^S$. Let $\hat{\mathcal{J}} = \mathcal{J}(G \times S)^S / \mathcal{I}$, and let $\mathcal{M}(N, M, S, T)$ denote the maximal ideal in $\mathcal{F}(N, M, S, T)$. We have

Lemma 4.1 *$\hat{\mathcal{J}}$ is the $\mathcal{F}(N, M, S)$-module generated by V_1, V_4, V_5, V_6, and V_7. The complement of $\mathcal{J}(G \times S)^S$ in $\mathcal{M}(N, M, S, T) \times \mathcal{M}(N, M, S, T)$ equals the complement of $\hat{\mathcal{J}}$ in $\mathcal{M}(N, M, S) \times \mathcal{M}(N, M, S)$.*

A careful consideration of the remaining generators gives

Lemma 4.2 *If $\mathcal{M}(N, M, S) \times 0 \subset \mathcal{J}$ then $\mathcal{M}(N, M, S, T) \times \mathcal{M}(N, M, S, T) \subset \mathcal{J}$.*

For the tangent space one can prove a similar result.

Lemma 4.3 *If $\mathcal{M}(N, M, S) \times \mathcal{M}(G, S) \subset T\mathcal{A}_S(G \times S)$ then $\mathcal{M}(N, M, S, T) \times \mathcal{M}(N, M, S, T) \subset T\mathcal{A}_S(G \times S)$.*

This lemma allows us to organize the computations in such way that they involve the first component of the vector fields only. Note that among the polynomials N, M, S, T we have the relation

$$4MN = S^2 + T^2. \tag{16}$$

This relation allows us to rewrite $4MN$ as S^2 in $\hat{\mathcal{J}}$. Thus

Lemma 4.4 *In $\hat{\mathcal{J}}$ we have $\mathcal{M}(N, M, S) = \mathcal{M}(M, S) + \mathcal{M}(N, S)$.*

Let $U(N, M, S)$ be the projection onto the first component of the subspace of $\hat{\mathcal{J}} + \mathcal{M}(G, S) \times \mathcal{M}(G, S)$ which contains those maps with zero second component. To obtain $U(N, M, S)$ it is sufficient to take in $\hat{\mathcal{J}}$ the generators with second component zero, S or G because one can consider $\hat{\mathcal{J}}$ as an $\mathcal{F}(G, S)$-module. Thus

Lemma 4.5

$$\begin{aligned} U(N, M, S) = &\mathcal{F}(N, M, S)\{N - aSM - bM^3\} + \\ &\mathcal{F}(G, S)\{2N + aSM\} + \\ &\mathcal{F}(G, S)\{15aS^2 + (8b + 12a^2)SM^2 + 4abM^4\} + \\ &\mathcal{M}(G, S). \end{aligned} \tag{17}$$

Let $U(N, S)$ and $U(M, S)$ denote the projections of $U(N, M, S)$ on $\mathcal{M}(M, S)$ and $\mathcal{M}(N, S)$ respectively. Clearly because of the first and last terms in $U(N, M, S)$ we have $U(N, S) = \mathcal{M}(N, S)$. Consequently the above lemmas combine to the following

Theorem 4.1 *The complement of $T\mathcal{A}_S(G \times S)$ in $\mathcal{M}(N, M, S, T) \times \mathcal{M}(N, M, S, T)$ equals the complement of $U(M, S)$ in $\mathcal{M}(M, S)$.*

The last theorem simplifies the computations necessary to obtain the codimension and unfolding of the map $G \times S$ considerably.

5 Topological unfolding of $G \times S$

Using the results of the previous section straightforward computations give

Theorem 5.1 *The complement of $T\mathcal{A}_{\mathcal{S}}(G \times S)$ in $\mathcal{M}(N,M,S,T)\times\mathcal{M}(N,M,S,T)$ is $< M, MS, MS^2, M^2 >$ (considered as $\mathbb{R}$-module) provided $a^2+b \neq 0$, $b-3a^2 \neq 0$, $b \neq 0$.*

Consequently the $\mathcal{A}_{\mathcal{S}}$-codimension of $G \times S$ is 4 and the $\mathcal{A}_{\mathcal{S}}$-minimal-universal-unfolding is $\bar{G}_\lambda \times S$ with

$$\bar{G}_\lambda = N + aMS + \lambda_1 M + \frac{\lambda_2}{2}M^2 + \frac{b}{3}M^3 + \lambda_3 MS + \lambda_4 MS^2. \tag{18}$$

Note that our choice of G was rather arbitrary. With

$$\tilde{G} = N + aMS + bM^3 + cMS^2, \tag{19}$$

we find that $\tilde{G} \times S$ has $\mathcal{A}_{\mathcal{S}}$-codimension 3 and unfolding terms $\{M, MS, M^2\}$. Futhermore with

$$\hat{G} = N + bM^3 + cMS^2, \tag{20}$$

we find that $\hat{G} \times S$ has $\mathcal{A}_{\mathcal{S}}$-codimension 3 and unfolding terms $\{M, MS, M^2\}$. It is clear that $G \times S$, $\tilde{G} \times S$, and $\hat{G} \times S$ are not $\mathcal{A}_{\mathcal{S}}$-equivalent. We can get around this problem of choice by the observation that for the description of bifurcations it is sufficient to consider topological $\mathcal{A}_{\mathcal{S}}$-equivalence in order to distinguish between qualitatively different bifurcation diagrams. In the papers Damon [1988a] and Damon [1988b] the Thom-Mather theory of C^∞-equivalence is connected with the theory of topological equivalence (using homeomorphisms instead of diffeomorphisms). The following theorems are taken from Damon [1988a] and Damon [1988b], but the formulation is adapted to the situation at hand. For sake of convenience we will talk about unfoldings of G instead of $G \times S$ because it is clear that we only unfold in the first component of the mapping.

On $\mathbb{R}^4$ we may assign a degree to each coordinate thus obtaining a grading on the space of polynomials on $\mathbb{R}^4$.

Theorem 5.2 *Given a grading δ. If G is δ-homogeneous and of finite $\mathcal{A}_{\mathcal{S}}$-codimension, then any unfolding of nondecreasing δ-degree is a topologically trivial unfolding.*

We call an unfolding universal of δ-degree $< m$ if we only have the unfolding terms of δ-degree $< m$ in the minimal universal $\mathcal{A}_{\mathcal{S}}$-unfolding.

Theorem 5.3 *If G is δ-homogeneous then a universal unfolding of G of δ-degree $< \delta(G)$ is a topologically universal unfolding.*

Note that in the situation of the degenerate Hamiltonian Hopf bifurcation any grading chosen should respect the relation (16). Furthermore it should make the lower order part of $\bar{H}$ without the terms that only depend on S, namely G, homogeneous. This gives us the conditions $\delta(N) = \delta(MS) = 3\delta(M)$, and $\delta(N) + \delta(M) = 2\delta(S)$. We obtain, up to a multiplicative constant, the unique grading

$$\delta(M) = 2, \delta(N) = 6, \delta(S) = 4. \tag{21}$$

Thus according to the above theorems the unfolding terms MS^2, MS are topologically redundant. Furthermore G and $\tilde{G}$ are topologically equivalent. Topologically we still have to distinguish between $a = 0$ and $a \neq 0$. We can get around this problem by considering a as a parameter. We obtain the following topologically minimal universal $\mathcal{A}_S$-unfolding

$$G_\lambda = N + aMS + \lambda_1 M + \frac{\lambda_2}{2}M^2 + \frac{b}{3}M^3, \tag{22}$$

provided

$$a^2 + b \neq 0, b - 3a^2 \neq 0, b \neq 0. \tag{23}$$

6 Analysis of the bifurcation

In order to get the bifurcation diagrams for the degenerate Hamiltonian Hopf bifurcation as introduced in section 1 we now have to analyse the Hamiltonian system corresponding to the singularity theoretic normal form for the bifurcation given by (22). Solutions of the Hamiltonian vector field corresponding to S have the form $z(t) = R_t\tilde{z}, \tilde{z} \in I\!R^4$, with $R_t(\tilde{x},\tilde{y}) = (e^{Jt}\tilde{x}, e^{Jt}\tilde{y})$, and

$$J = \begin{pmatrix} 0 & -1 \\ 1 & 0 \end{pmatrix}. \tag{24}$$

For sufficiently small (λ,σ) all sufficiently small $2\pi(1+\sigma)^{-1}$ periodic solutions of the Hamiltonian vector field X_{G_λ} corresponding to G_λ are of the form $z(t) = R_{(1+\sigma)t}\tilde{z}$. Apply to Hamilton's equations corresponding to the Hamiltonian G_λ the transformation $z = R_{(1+\sigma)t}\tilde{z}$. We get

$$\begin{aligned} \dot{\tilde{x}} &= -(\sigma - aM)J\tilde{x} + (aS + bM^2 + \lambda_2 M + \lambda_1)\tilde{y}, \\ \dot{\tilde{y}} &= -(\sigma - aM)J\tilde{y} - \tilde{x}. \end{aligned} \tag{25}$$

All stationary points of (25) correspond to $2\pi(1+\sigma)^{-1}$ periodic solutions of X_{G_λ}. Note that the right hand side of the above equations equals $\mathrm{d}G_\lambda$-$\sigma\mathrm{d}S$. The equation $\mathrm{d}G_\lambda = \sigma\mathrm{d}S$ gives the set of critical points of the map $G_\lambda \times S$, i.e. the bifurcation equation. Put $\alpha = \sigma - aM$, and suppose $M > 0$, then the stationary points of (25) are given by the equations

$$\begin{aligned} \tilde{x} &= -\alpha J\tilde{y}, \\ \alpha^2 &= aS + bM^2 + \lambda_2 M + \lambda_1. \end{aligned} \tag{26}$$

The first equation in (26) gives furthermore that along the periodic solutions

$$S = 2\alpha M, N = \alpha^2 M, T = 0. \tag{27}$$

Note that for $M = 0$ the only stationary point is the origin. Consequently the equilibria (curve of periodic solutions) are given by

$$CPS : \alpha^2 = 2a\alpha M + \lambda_2 M + bM^2 + \lambda_1, \tag{28}$$

Straightforward linearization of the equations (25) gives us for the stability:

(i) elliptic if $2\alpha^2 + 2bM^2 + \lambda_2 M > 0$,

(ii) hyperbolic if $2\alpha^2 + 2bM^2 + \lambda_2 M < 0$.

Define the curve of stability transitions

$$CST : 2\alpha^2 + 2bM^2 + \lambda_2 M = 0. \tag{29}$$

The bifurcation diagrams can now be classified according to the number of intersections of the CPS and CST which lie in the $M > 0$ halfplane. The CPS is an ellipse if $(a^2 + b) < 0$ and $\left(4\lambda_1 - \frac{\lambda_2^2}{(a^2+b)}\right) > 0$ which degenerates into a point if $\left(4\lambda_1 - \frac{\lambda_2^2}{(a^2+b)}\right) = 0$ and becomes empty if $\left(4\lambda_1 - \frac{\lambda_2^2}{(a^2+b)}\right) < 0$. The CPS is an hyperbola if $(a^2 + b) > 0$ which degenerates into two lines if $\left(4\lambda_1 - \frac{\lambda_2^2}{(a^2+b)}\right) = 0$. The CST is an ellipse if $b > 0$ and a hyperbola if $b < 0$. The center of the CPS lies on the CST. We now have to study the intersections of these quadrics in dependance of the parameters b, a, λ_1, and λ_2. Note that when a changes sign the picture is reflected in the M-axis and that a sign change of λ_2 gives a reflection in the α-axis.

The following computations give the precise conditions for the different cases we have to distinguish. First eliminate α from the CPS and CST to obtain Res_M. Also eliminate M from the CPS and CST to obtain Res_α. The zeroes of Res_M and Res_α give the coordinates of the intersections of CPS and CST. We have

$$\begin{aligned} Res_M =& 16b(b + a^2)M^4 + 8\lambda_2(3b + a^2)M^3 + \\ & (9\lambda_2^2 + 16b\lambda_1)M^2 + 12\lambda_1\lambda_2 M + 4\lambda_1^2, \end{aligned} \tag{30}$$

$$\begin{aligned} Res_\alpha =& 16b(b + a^2)\alpha^4 + 16ab\lambda_2\alpha^3 - b(16b\lambda_1 - 3\lambda_2^2)\alpha^2 - \\ & 4ab\lambda_1\lambda_2\alpha - b\lambda_1\lambda_2^2 + 4b^2\lambda_1^2. \end{aligned} \tag{31}$$

Let Δ_M and Δ_α denote the discriminants of Res_M and Res_α. We have

$$\begin{aligned} \Delta_M =& a^4 b \left(a^2 + b\right) {\lambda_1}^3 \left(4(a^2 + b)\lambda_1 - {\lambda_2}^2\right) \\ & \left(1024b^3{\lambda_1}^2 - 576b^2\lambda_1{\lambda_2}^2 - 27(a^2 - 3b){\lambda_2}^4\right), \end{aligned} \tag{32}$$

$$\begin{aligned} \Delta_\alpha =& b^7 \left(a^2 + b\right) \lambda_1 \left(8a^2\lambda_1 - {\lambda_2}^2\right)^2 \left(4(a^2 + b)\lambda_1 - {\lambda_2}^2\right) \\ & \left(1024b^3{\lambda_1}^2 - 576b^2\lambda_1{\lambda_2}^2 - 27(a^2 - 3b){\lambda_2}^4\right). \end{aligned} \tag{33}$$

The zeroes of Δ_M and Δ_α give us the points where the number of M-coordinates, respectively, the number of α-coordinates, of the intersection points of CPS and CST change. Thus changes in the number of intersections of CPS and CST only occur when $\Delta_M = \Delta_\alpha = 0$, i.e. $\Delta_M = \Delta_\alpha = 0$ give us conditions on the parameters a, b, λ_1, and λ_2 which determine the different cases we have to consider.

The conditions on a and b coincide with those following from the singularity theory and are given in Figure 1. The conditions divide the (a, b)-plane into several regions. When we restrict to $a > 0$ we obtain the following four cases:

I $a > 0,\ b > 3a^2$.

II $a > 0,\ 0 < b < 3a^2$.

III $a > 0,\ -a^2 < b < 0$.

IV $a > 0,\ b < -a^2$.

Recall that for $a < 0$ the pictures in the (α, M)-plane are reflected in the M-axis, while for $a = 0$ the pictures are symmetric with respect to the M-axis. The conditions on λ_1 and λ_2 following from $\Delta_M = \Delta_\alpha = 0$ in the four cases I, II, III, and IV

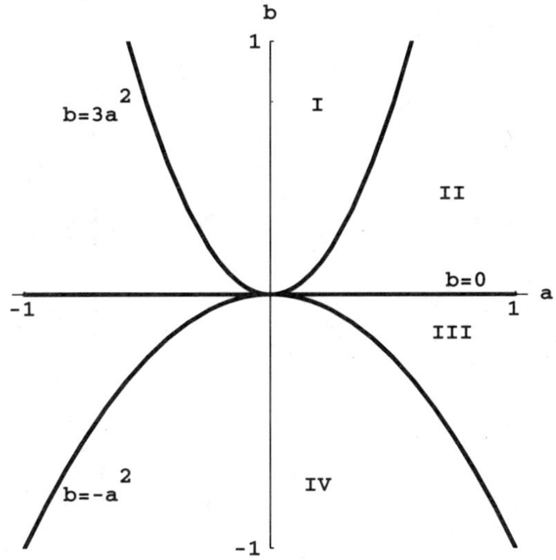

Figure 1 Conditions on a and b.

are given in Figures 2, 3, 4, 5. The factor $\left(1024b^3{\lambda_1}^2 - 576b^2\lambda_1{\lambda_2}^2 - 27(a^2-3b){\lambda_2}^4\right)$ only has zeroes if $b > 0$. Besides $\lambda_1 = 0$ we have to consider the following curves in the (λ_1, λ_2)-plane:

$$
\begin{aligned}
c_1 : \lambda_1 &= \frac{1}{4(a^2+b)}\lambda_2^2, \\
c_2 : \lambda_1 &= \frac{3\sqrt{3}}{32b\sqrt{b}}(\sqrt{3}\sqrt{b}+a)\lambda_2^2, \\
c_3 : \lambda_1 &= \frac{3\sqrt{3}}{32b\sqrt{b}}(\sqrt{3}\sqrt{b}-a)\lambda_2^2.
\end{aligned} \tag{34}
$$

The curves c_2 and c_3 only play a role in the cases I and II. Because we only consider intersections of CPS and CST in the $M > 0$ halfplane, we only consider the part of the curves c_1, c_2 and c_3 in the $\lambda_2 < 0$ halfplane in these two cases. In the cases II and IV we only have to consider c_1, we restrict the curve to the halfplane $\lambda_2 < 0$, $\lambda_2 > 0$ respectively. Note that at $b = 3a^2$ the curves c_1 and c_3 exchange places.

The Figures 2, 3, 4, and 5 give diagrams from which it is clear which cases we have to consider. The actual bifurcations take place at the thickened curves. For each of the numbered area's or curves we will give the corresponding situation in the (α, M)-plane in Figures 6, 7, 8, and 9. The thickened curve is the CPS. The grey area denotes the area in which the equilibria are unstable, its boundary is the CST.

In Figure 6 we consider case I. The CPS is a hyperbola and the CST an ellipse, which only lies in the $M > 0$ halfplane for $\lambda_2 < 0$. In (I.1) we have two stable branches of equilibria attached to the α-axis. Along c_2, (I.2), along one of the branches an unstable part comes into existence. Along c_3, (I.4), the same happens along the other branch. Along c_1, (I.6), the hyperbola degenerates into two lines, i.e. the two branches become attached to each other. After this we again have two branches, (I.7). One of the branches contains an unstable part and is attached to the α-axis, the other is a stable branch which lies away from the α-axis. Along the negative λ_2-axis, (I.8), the branch with the unstable part pulls into the origin. The

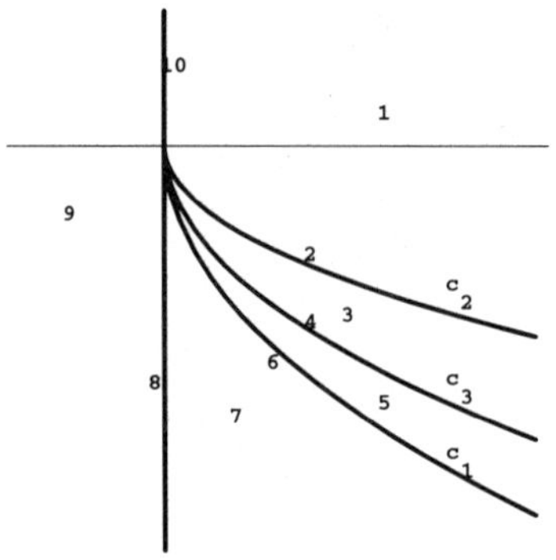

Figure 2 (λ_1, λ_2)-plane in case I.

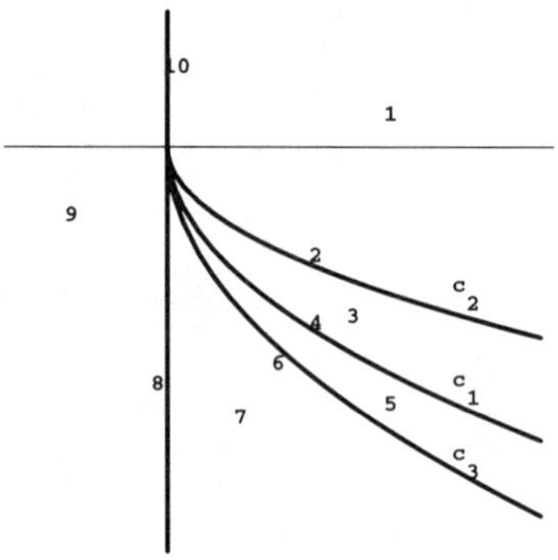

Figure 3 (λ_1, λ_2)-plane in case II.

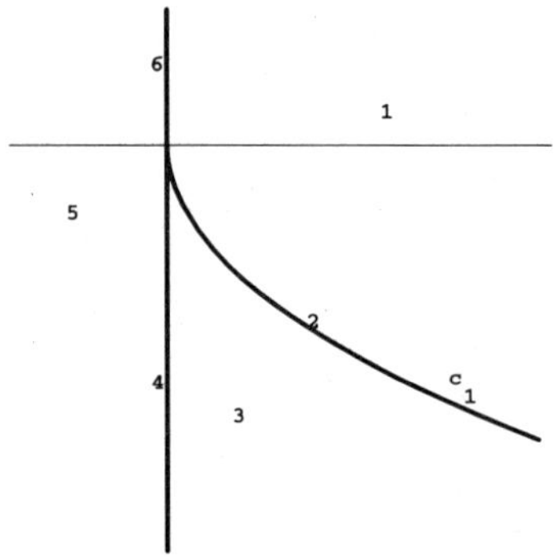

Figure 4 (λ_1, λ_2)-plane in case III.

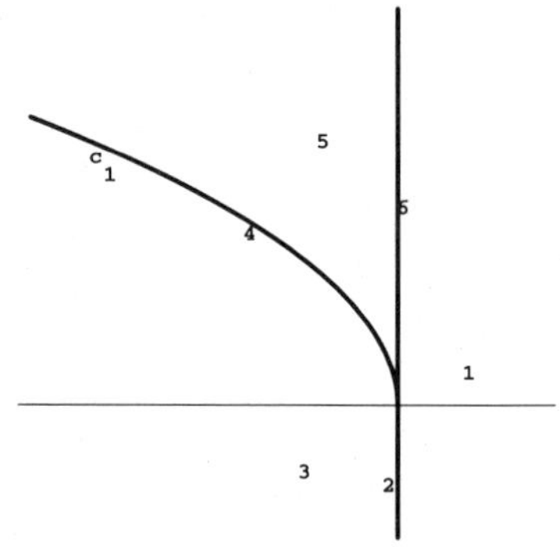

Figure 5 (λ_1, λ_2)-plane in case IV.

remaining branch now becomes attached to the origin (along the positive λ_2-axis, I.10) and splits into two stable branches, which completes the cycle.

In Figure 7 we consider case II. Again the CPS is a hyperbola and the CST an ellipse, which only lies in the $M > 0$ halfplane for $\lambda_2 < 0$. Like in case I we start, (II.1), with two stable branches of equilibria attached to the α-axis. Along c_2, (I.2), along one of the branches an unstable part comes into existence. Along c_1, (II.4), the hyperbola degenerates into two lines, i.e. the two branches become attached to each other. After this we again have two branches (II.5), now both contain an unstable part, one branch is attached to the α-axis, the other lies away from the α-axis. Along c_3, (II.6), on the branch away from the α-axis the unstable part dissapears, which gives us two branches (II.7), now one contains an unstable part and is attached to the α-axis, the other is a stable branch which lies away from the α-axis. Along the negative λ_2-axis, (II.8), the branch with the unstable part pulls into the origin. The remaining branch now becomes attached to the origin (along the positive λ_2-axis, II.10) and splits into two stable branches, which completes the cycle.

In Figure 8 we consider case III. Both the CPS and the CST are hyperbola. We start, (III.1), with two stable branches of equilibria attached to the α-axis. One of the branches is unstable except for a part close to the α-axis. Along c_1, (III.2), the CPS-hyperbola degenerates into two lines, i.e. the two branches become attached to each other. After this we again have two branches (III.3), now both contain an

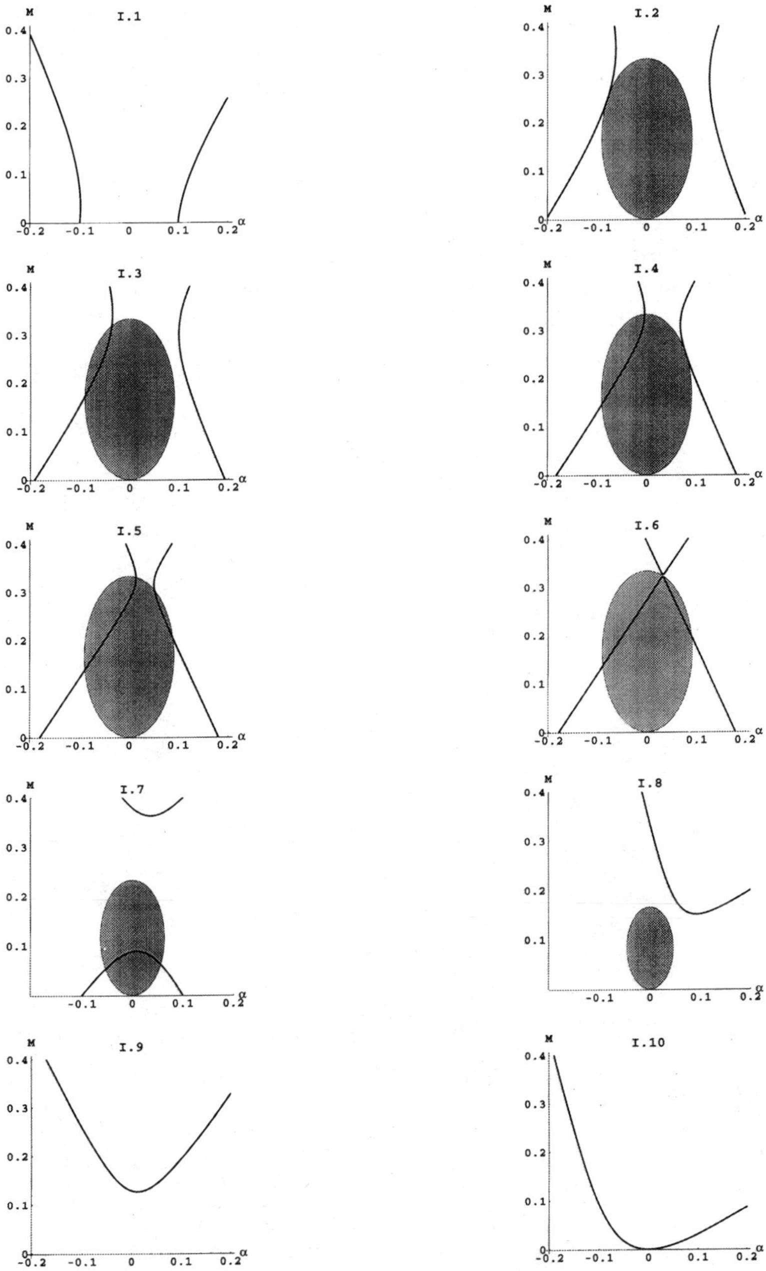

Figure 6 CPS and CST in case I.

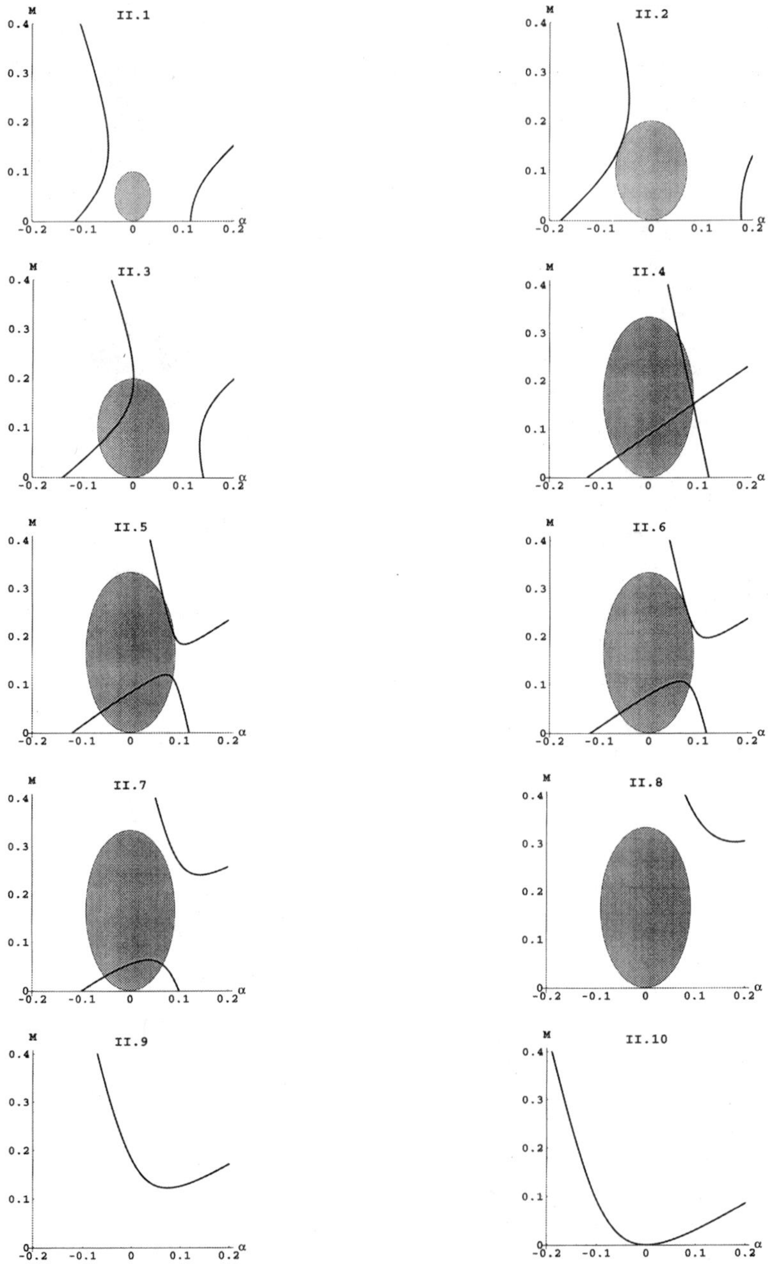

Figure 7 CPS and CST in case II.

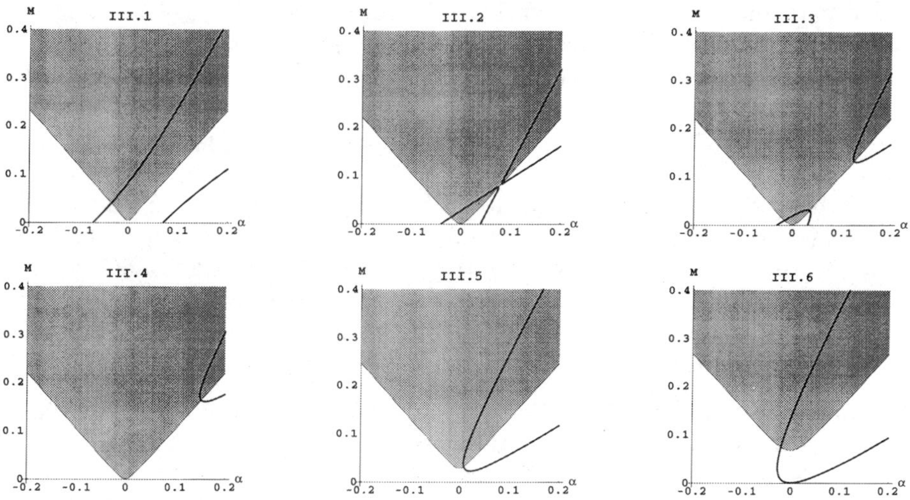

Figure 8 CPS and CST in case III.

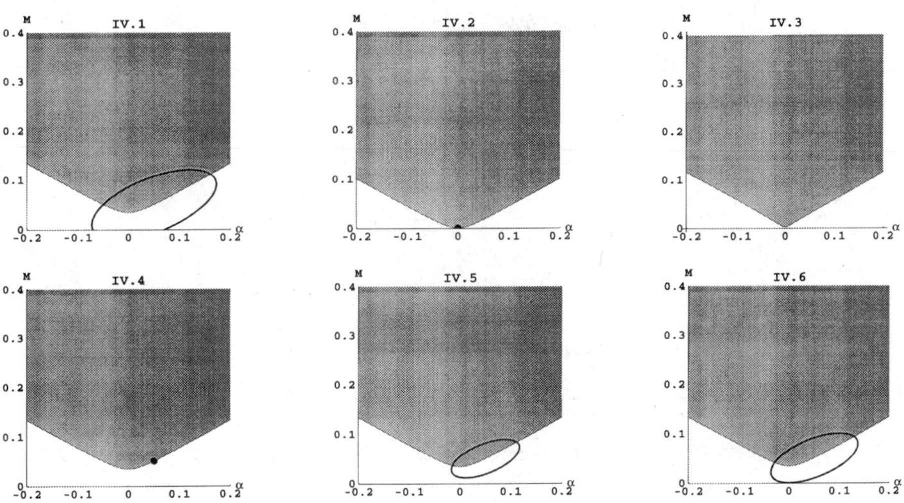

Figure 9 CPS and CST in case IV.

unstable part, one branch is attached to the α-axis, the other lies away from the α-axis. Along the negative λ_2-axis, (III.4), the branch attached to the α-axis pulls into the origin, the remaining part away from the α-axis still contains an unstable part. This branch now becomes attached to the origin (along the positive λ_2-axis, III.6) and splits into two branches, one with an unstable part, which brings us back to our starting point.

In Figure 9 we consider case IV. The CPS is an ellipse and the CST a hyperbola. We start, (IV.1), with a branch containing an unstable part and at both sides attached to the α-axis. Along the negative λ_2-axis, (IV.2), this branch pulls into the origin. Along c_1, (IV.4), out of the blue comes a closed curve of equilibria part of which is unstable. Along the positive λ_2-axis, (IV.6), this curve becomes attached to the origin and breaks open along the α-axis.

Remark In Van der Meer [1985] the bifurcation was analyzed by factoring the energy-momentum mapping $G_\lambda \times S$ through the orbit map for the $\mathcal{S}$-action. This way one obtains the periodic solutions as the critical points of the reduced Hamiltonian on the reduced phase spaces which are given by the relation (16) together with $S = \textit{constant}$. The equations in M, N, S, and T one obtains this way are equivalent to (27)and (28) together with the Hamiltonian. Thus from these equations one can obtain a description of the critical values of the energy-momentum mapping in the (S, G)-plane like in Van der Meer [1985]. Precise pictures will be published elsewhere.

Remark Recall the equations for the equilibria (at $\lambda_1 = \lambda_2 = 0$)

$$\begin{aligned} 0 &= -(\sigma - aM)J\tilde{x} + (aS + bM^2)\tilde{y}, \\ 0 &= -(\sigma - aM)J\tilde{y} - \tilde{x}. \end{aligned} \tag{35}$$

If $M \neq 0$, then this is equivalent to

$$\begin{aligned} \sigma S &= 3aSM + 2bM^3, \\ 2\sigma M &= S + 2aM^2, \\ 4MN &= S^2, \\ T &= 0. \end{aligned} \tag{36}$$

Note that if one puts $\alpha = \sigma - aM$ then equations (36) are equivalent to (27) and (28). Eliminate S using the first two equations of (36). This gives an equation in M and σ, $M > 0$. Put $M = x^2$, then we get a $\mathbb{Z}_2$ invariant equation. We get $(b - 3a^2)M^2 + 4a\sigma M - \sigma^2 = 0$. If we set $M = x^2$, $\lambda = \frac{\sigma}{\sqrt{(b-3a^2)}}$, $m = \frac{2a}{\sqrt{(b-3a^2)}}$, we get

$$\delta x^4 + 2m\lambda x^2 - \lambda^2 = 0, \tag{37}$$

with $\delta = 1$ if $b - 3a^2 > 0$, and $\delta = -1$ if $b - 3a^2 < 0$. Note that equation (37) is precisely the normal form of which the unfolding is analysed by Bridges in Bridges [1990] and Bridges [1991] to obtain the qualitative description for the degenerate Hamiltonian Hopf bifurcation. Our results coincide with his results. Actually equation (37) is precisely one of the singularity theoretic topological normal forms obtained in the classification of $\mathbb{Z}_2$ symmetric singularities in Golubitsky and Schaeffer [1985].

References

Bierstone, E. [1980], *The structure of orbit spaces and the singularities of equivariant mappings*, Monografias de Matematica **35**, Inst. de Mat. Pura et Applicada, Rio de Janeiro.

Bridges, T.J. [1990], *Bifurcation of periodic solutions near a collision of eigenvalues of opposite signature*, Math. Proc. Cambridge Philos. Soc., **108**, 575-601.

Bridges, T.J. [1991], *Stability of periodic solutions near a collision of eigenvalues of opposite signature*, Math. Proc. Cambridge Philos. Soc. **109**, 375-403.

Burgoyne, N. and Cushman, R.H. [1974], *Normal forms for real linear Hamiltonian systems with purely imaginary eigenvalues*, Celestial Mech., **8**, 435-443.

Cushman, R.H., Deprit, A. and Mosak, R. [1983], *Normal form and representation theory*, J. Math. Phys. **24**, 2103-2116.

Cushman, R.H. and Sanders, J.A. [1987], *Nilpotent normal forms and representation theory of* $sl(2, \mathbb{R})$, in: Multiparameter bifurcation theory, eds. M. Golubitsky, J. Guckenheimer, Contemp. Math., **56**, Amer. Math. Soc., Providence, R.I., 31-51.

Cushman, R.H. and Sanders, J.A. [1988], *Splitting algorithm for nilpotent normal forms*, Dynamics Stability Systems, **2**, 235-246.

Cushman, R.H. and Sanders, J.A. [1990], *A survey of invariant theory applied to normal forms of vectorfields with nilpotent linear parts*, in: Invariant theory and tableaux, Minneapolis 1988, ed. D. Stanton, IMA Vol. Math. Appl., **19**, Springer-Verlag, Berlin, 82-106.

Cushman, R.H. and Van der Meer, J.C. [1990], *The Hamiltonian Hopf bifurcation in the Lagrange top*, in: Géométrie Symplectique et Mécanique, V Colloque International du Séminaire Sud-Rhodanien de Géométrie, 1988, ed. C. Albert, Lecture Notes in Math., **1416**, Springer-Verlag, Berlin, 26-38.

Damon,J.N. [1988a], *Topological triviality and versality for subgroups of* $\mathcal{A}$ *and* $\mathcal{K}$, Mem. Amer. Math. Soc., **389**, Amer. math. Soc., Providence, R.I.

Damon,J.N. [1988b], *Topological equivalence of bifurcation problems*, Nonlinearity **1**, 311-331.

Duistermaat, J.J. [1984], *Bifurcations of periodic solutions near equilibrium points of Hamiltonian systems*, in: Bifurcation theory and applications, Montacatini 1983, (L. Salvadori, ed.) Lecture Notes in Math., **1057**, Springer-Verlag, Berlin, 55-105.

Golubitsky, M. and Guillemin, V. [1973], *Stable mappings and their singularities*, Graduate Texts in Math., **14**, Springer-Verlag, New York.

Golubitsky, M. and Schaeffer, D.G. [1985], *Singularities and groups in bifurcation theory, vol. 1*, Appl. Math. Sci., **51**, Springer-Verlag, New York.

Marsden, J.E. [1992], *Lectures on mechanics*, London Math. Soc. Lecture Note Ser., **174**, Cambridge Univ. Press, Cambridge.

Martinet,J. [1982], *Singularities of smooth functions and maps*, London Math. Soc. Lecture Note Ser., **58**, Cambridge Univ. Press, Cambridge.

Van der Meer, J.C. [1985], *The Hamiltonian Hopf bifurcation*, Lecture Notes in Math., **1160**, Springer-Verlag, Berlin.

Van der Meer, J.C. [1990], *Hamiltonian Hopf bifurcation with symmetry*, Nonlinearity, **3**, 1041-1056.

Poènaru, V. [1976], *Singularités C^∞ en présence de symétrie*, Lecture Notes in Math., **510**, Springer-Verlag, Berlin.

Roberts, M. [1986], *Characterisations of finitely determined map germs*, Math. Ann., **275**, 583-597.

Schwarz, G. [1975], *Smooth functions invariant under the action of a compact Lie group*, Topology, **14**, 63-68.

Vanderbauwhede, A and Van der Meer, J.C.: [1995], *A general reduction principle for periodic solutions in Hamiltonian systems*, in: Normal Forms and Homoclinic Chaos, (W.F. Langford and W. Nagata, eds.), Fields Institute Communications, **4**, AMS, Providence, R.I., 273-294.

Williamson, J. [1936], *On the algebraic problem concerning the normal forms of linear dynamical systems*, Amer. J. Math., **58**, 141-163.